THE UPS AND DOWNS
OF PHYSICS

JUDITH WEBER
AND **MARCUS WEBER**

Translated by
ELIZABETH SCHWAIGER

THE UPS & DOWNS OF PHYSICS

The Science of Gravity, Aerodynamics, and Everyday Mishaps

GREYSTONE BOOKS
Vancouver/Berkeley/London

Copyright © 2025 by Marcus and Judith Weber
First published in English by Greystone Books in 2025
Originally published in German as *Phantastisch physikalisch*,
Marcus and Judith Weber copyright © 2021 by Wilhelm Heyne Verlag
English translation copyright © 2025 by Elizabeth Schwaiger
Illustrations copyright © 2025 by Marcus Weber

25 26 27 28 29   5 4 3 2 1

The publisher expressly prohibits the use of *The Ups and Downs of Physics* in connection with the development of any software program, including, without limitation, training a machine-learning or generative artificial intelligence (AI) system.

All rights reserved, including those for text and data mining, AI training, and similar technologies. No part of this book may be reproduced, stored in a retrieval system or transmitted, in any form or by any means, without the prior written consent of the publisher or a license from The Canadian Copyright Licensing Agency (Access Copyright). For a copyright license, visit accesscopyright.ca or call toll free to 1-800-893-5777.

Greystone Books Ltd.
greystonebooks.com

Cataloguing data available from Library and Archives Canada
ISBN 978-1-77840-064-3 (cloth)
ISBN 978-1-77840-065-0 (ebook)

Copy editing by Sarah Pybus
Proofreading by Dawn Loewen
Indexing by Stephen Ullstrom
Jacket and text design by Fiona Siu
Jacket images by nurulanga/iStock.com (sofa); Boonyachoat/iStock.com (sky); Cheer Group/Shutterstock.com (apples)
Printed and bound in Canada on FSC® certified paper at Friesens. The FSC® label means that materials used for the product have been responsibly sourced.

Greystone Books thanks the Canada Council for the Arts, the British Columbia Arts Council, the Province of British Columbia through the Book Publishing Tax Credit, and the Government of Canada for supporting our publishing activities.

Canadä

Greystone Books gratefully acknowledges the xʷməθkʷəy̓əm (Musqueam), Sḵwx̱wú7mesh (Squamish), and səlilwətaɬ (Tsleil-Waututh) peoples on whose land our Vancouver head office is located.

Contents

1. Introduction, or How to Extinguish a Salmon on Fire *1*

2. Superman on a Bike *3*

3. Toast, Tires, and Space *18*

4. Hello? Hello? Are You Still There? *34*

5. A Collapsing Bridge? That's So Boring! *53*

6. Never Throw a Sofa out a Window *65*

7. The Greenhouse Effect in Children's Rooms *83*

8. High-Rise Melts Car Parts *97*

9. Lazy Bee Against a Blue Sky *106*

10. Electricity Hurts *118*

11. Fingernails on a Blackboard *135*

12. Foggy Glasses and Cloudy Mirrors *149*

13. RIP, Cell Phone! *162*

14. What a Beautiful Glow! *174*

15. Pistol Shrimp and Sinking Ships *192*

Thank you *207*

Index *209*

Introduction, or How to Extinguish a Salmon on Fire

"HERE'S WHAT I KNOW about physics: Things fall and electricity hurts!" We have this postcard on our office wall, and its message is true. Physics can really get on your nerves, so much so that sometimes you never want to hear about it ever again. Take the barbecue when the salmon caught fire. It was a warm summer evening, and we were with friends in their backyard. We cracked open the first beer and smelled the wonderful aroma of a salmon sprinkled with herbs. We said "Cheers," and then we screamed: Flames were shooting up from the grill. Marcus's first thought was "Water!" Or should we have just doused it with our beers?

Luckily the hosts knew more about the physics of grilling than the only physicist at the gathering. They stopped Marcus and swiftly rescued the salmon from the grill with long barbecue tongs. As the flames flickered out, they laughed long and hard at how the only one in the group who had actually studied physics had the reflexive instinct to throw water onto a grease fire. Because the result would have been a spectacular ball of fire. The water (or beer) would have immediately evaporated on the glowing grill and the vapor would have picked up countless small droplets of fat, which would have burst into flames. The surface area of burning fat would have been greatly increased. Thanks, physics!

Physics makes life hard for us in various situations. It does its thing whether we like it or not. Riding a bike, there will always be headwinds; glasses fog up; and mobile Wi-Fi connections drop.

Now let's get to the "but." No postcard slogan tells the other side of the story—namely, that each silly mishap and each annoying effect is underpinned by a beautiful, elegant physical principle. A natural principle that helps us through life in other instances. So let's go and look at all the times when physics has made life difficult for us. And let's find out why. Then we're going to turn it all inside out, or upside down. For when the right tricks are applied, physics works in our favor. In those instances, we take advantage of its effects, and the headwind on our way home feels like a fresh breeze, spurring our brains on to peak performance. It's a promise! Enjoy!

Superman on a Bike

Nothing but headwinds and how to conquer them

Cycling tours always look relaxing in travel brochures. Smiling people ride through dreamy landscapes; the sun is shining, the meadows are in bloom, and their hair is lifted elegantly by a light breeze. Our holiday pictures tell a different story. There we are, bent over the handlebars and pedaling hard, our faces bright red with exertion, our T-shirts fluttering—inelegantly—in the wind. The album of our first holiday is full of such pictures. We toured Cuba on bikes for four weeks. Fidel Castro was still alive and our children were as yet unborn—it was perfect timing. We checked our bicycles as oversized luggage at Frankfurt Airport, retrieved them in Havana, and set off. We encountered many challenges over the course of those four weeks and found solutions for most of them:

- Struggling to buy food? Bananas grow along the verges and a whole stalk on the bike rack doesn't really get in the way of riding.

- Not allowed to pitch a tent? There are always nice people ready to offer a sofa for the night—as long as we leave the house before daybreak, so the police won't notice.

- Few people speak English? Just "spanify" French by changing the emphasis and adding an "o" to as many words as possible—it's amazing how well this works.

Just one problem remained: headwinds. It didn't matter whether we were cycling along the coast, through the interior, or past mountains, whether we were cycling eastward, southward, or to the north; the wind was not our friend. As long as the route was full of interest and variation, it didn't matter. There was so much to see. But after we'd spent hours one day struggling along a gravel road leading to nowhere, there was only one topic of conversation that evening: Is there no alternative? Isn't it possible to cycle without constant headwinds? There must be a way to conquer these pesky winds—or even to make use of them!

The next day, we tried a small experiment. From now on we would pay close attention every morning to where the wind was blowing from before getting on our bikes. Perhaps there were wind directions that didn't result in headwinds? Or at least lighter headwinds? But we were riding along the coast and, generally speaking, the wind was coming from the sea. These observations were of little help.

But then there was a day with virtually no wind. The sea was like a mirror and the blades of grass along the path were still. Hurrah! Finally, a day without headwinds! Highly motivated, we mounted our bikes, set off, and felt... headwinds. They weren't light either. It's only logical: When we move forward, the airflow blows in our faces. We are riding against the air and have to push through it, so to speak. Still, we were surprised how strong the perceived headwinds were.

Upon our return, we threw the bikes into a corner and started to wrestle with the phenomenon of headwinds (it's always a good idea to know your enemies if you want to conquer them). We soon came to a frustrating realization: We were the problem. The bulk of the energy we exert while pedaling is expended to work against the air resistance created by our own body. Depending on body position and speed, this could be up to 90 percent. In other words, most of our energy is directed at fighting a self-created problem. Sometimes physics can be depressing!

Air is heavier than it seems

But it's no use, we must face facts: Under normal circumstances we don't really feel the air around us. It is simply there. Nevertheless, it is pressing on us and has quite a hefty weight. One cubic meter of air (thirty-five cubic feet) can weigh as much as 1.2 kilograms (2.6 pounds)! And when this mass is set in motion, we're in trouble. If we stand perfectly still in the middle of a meadow, we present an obstacle for the air that flows around us; we are simply in the way and the air wants to pass through. Let's assume there's a twenty-kilometer-per-hour (twelve-mile-per-hour) wind. For a person of average size, this would exert seven kilograms (fifteen pounds) of air per second. Let me repeat, per second! The larger the person, the greater the pressure. For the blades of a large wind turbine, fifty tons (100,000 pounds) of pressure flows across the area of each blade per second. This enormous mass provides a good sense of why wind turbine installations create so much electrical energy.

So we experience a "breaking" force even when there is no wind due to flow resistance or drag. For an average adult traveling at a speed of 20 km/h (12 mph), this resistance is roughly ten newtons (N or the unit of force). This is the force needed to hold (or lift) one kilo (2.2 pounds), the weight of a liter (0.26 gallons) of milk. This doesn't mean the equivalent of putting the milk in our bicycle basket; instead, what it means is that we are continuously dragging a liter of milk up via a thin rope and pulley, for example. This is the force we must expend to push through the air that lies in our path. And that's when there's no wind!

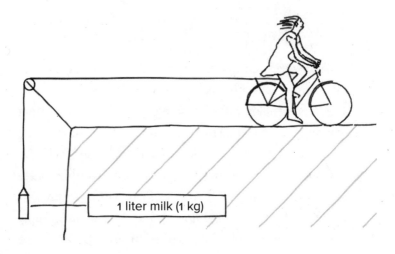

And the air doesn't make it easy, either at the back or at the front. At the back, this is because we are not built aerodynamically. As a cyclist, you are in effect an irregularly shaped body. This doesn't sound very flattering, but that's how it is from a physics perspective. An irregularly shaped body creates air eddies. These eddies dissolve and result in a small amount of negative pressure. You yourself create higher pressure in front of you because you are pushing through the air. This pressure gradient pulls you backward—mathematically speaking—which is why normal *airflow* feels like a headwind.

Add to that the actual wind—that is, the wind you feel when you take a well-deserved break (sailors speak of *true wind*). Combined, the airflow and the true wind result in *relative wind*.[1] This is the wind we feel when we're riding a bike and we have to pedal against it. When I'm cycling at a speed of 20 km/h (12 mph) and a true wind of 20 km/h is blowing against me at the same time, the result is a relative wind of 40 km/h (25 mph). Meteorologists call such wind speeds "strong winds," the kind that turn umbrellas inside out and set thick tree branches swaying.

When we read this, we felt like true heroes. In other words, we had cycled into officially strong winds every single day! We felt even more heroic when we reminded ourselves of the dirty physics trick that is flow resistance. For flow resistance increases by multiples the faster we cycle. Air resistance is nasty. It varies as the square of the speed of airflow. "Square" in this instance doesn't mean that it comes around four corners, but that it increases

---

1 Sailors also speak of "apparent wind."

disproportionately. If I cycle twice as fast, it quadruples. If I cycle three times as fast, I encounter nine times the resistance. And if I ride four times as fast, resistance is sixteen times higher. Practically speaking, this means that when I cycle at a speed of 20 km/h (12 mph) on a wind-still day, I need to exert a force of ten newtons. But now there's a wind as well, blowing at me at a speed of 20 km/h. In other words, the relative wind has doubled. But the force I need to exert is forty newtons, not just double but quadruple. That's a lot of milk cartons I have to pull up. I'd rather transport banana stalks across Cuba...

The wind at our backs—finally!

This sobering research brought us down a peg (or two). And then, last summer, we had an unexpected small victory. We cycled from Germany's Ruhr region to the North Sea all the way to the port of Dagebüll, Schleswig-Holstein, where the ferry to the idyllic island of Amrum docks. The route took us 550 kilometers (342 miles) to the north, with a constant southwesterly wind. It really propelled us forward and was so fierce that we couldn't even go surfing without falling off our boards when we took a break at Lake Dümmer, Lower Saxony. Take that, crosswind!

On our bikes it didn't really feel like a wind at our backs, but we noticed how easy it was to pedal and how easily we covered ground. This, too, was rooted in physics: When the wind blows from behind at 20 km/h (12 mph) and I'm cycling at a speed of 20 km/h, I don't feel any wind at all. The only thing that can slow my forward momentum is the roll resistance of the bike wheels.

The absolute best is to ride downhill with the wind at your back. On our tour to the North Sea, we took the 40 km/h (25 mph) challenge. Every day we tried to find at least one section of the route where we could reach this speed going downhill with the wind at our backs.

But before you get your bike out of storage and take off cycling north, we have a sobering fact to share with you. A lot of tailwind is needed before we actually perceive it as such. In Germany, the wind is usually too slow in comparison to the speed of travel. It simply doesn't negate the flow resistance. Let's take Hanover as an example, because it's perfectly located in the middle of the country. There, the average wind speed is 3 m/s (9.8 ft/s). That translates to 10.8 km/h (6.7 mph). For a tailwind like that to act as a boost, you would have to ride slower than 10.8 km/h.

Besides, wind is a localized event; there is more of it in the north of Germany than in the south. And that was the fate that befell us on the final stretch of our tour. We still had forty-five kilometers (twenty-eight miles) to go to reach our destination. Part of our "baggage" by this time was a shattered knee (we're not getting any younger), a borrowed bicycle with an uncomfortable saddle (our bike, too, wasn't getting any younger and had deserted us halfway through the tour), and the time pressure to catch the ferry. And then the wind turned. The tailwind we had barely noticed now blew right into our faces with full force. Before setting off on our tour, we had read that "headwinds eat the cyclist's strength for breakfast," and that's exactly how it was. We pedaled, we swore, we toiled. We stopped at a snack bar to drink over-sweetened cocoa, and then we swore and pedaled and toiled onward.

Are we sounding a bit plaintive? Probably. But allow us to make our case with numbers. On this particular day, there was a real North Sea wind. Wind force 7, which is a headwind of about 51–61 km/h (32–38 mph). Add to that our cycling speed—admittedly no longer very fast, but sometimes we still managed 10 km/h (6 mph). This adds up to a relative wind of 66 km/h (41 mph), which rounded up corresponds to one hundred newtons of flow resistance. To cycle against that resistance is akin to constantly pulling ten kilograms (twenty-two pounds) on the aforementioned rope or riding up a mountain with a 10 percent gradient. That's quite a lot, especially if the mountain road is forty-five kilometers (twenty-eight miles) long. It also doesn't help to know that the pros on the legendary Tour de France section toward Alpe d'Huez have to overcome a 15 percent gradient.

It would have been easier if we'd been riding recumbent bicycles instead of standard bikes. This saves a lot of energy, not least because you provide less contact surface for the wind. You could even take it to extremes and optimize the recumbent bicycle aerodynamically. The best way to achieve this is to shape it into a mix of a cigarette and a suppository. This streamlined form greatly reduces the flow resistance; given identical wind speed, recumbent bicycles optimized in this fashion present only a tenth of the resistance compared to a normal bicycle, never mind our own "packhorses" with bicycle bags front and back. This makes it possible to reach riding speeds over 140 km/h (87 mph) with muscle power alone. But we really wouldn't like to sit, or rather recline, on a "bicycle" like this optimized to achieve speed records. They are so completely encased that there isn't even a window. You can only see the road on a screen (making it unlikely that these models will be approved for road use any time soon).

Our only solution, therefore, was to bend over our handlebars as much as possible to reduce flow resistance at least a little bit. For a short while, we even tried to ride in the slipstre .m. You see it all the time on the Tour de France. The riders are close together, one behind the other, so that the one out front has to expend a lot of energy and those behind ride in his slipstream. There's only one problem: You need to ride very close together to benefit from it. Pile-ups are more or less guaranteed. And this concept doesn't account for traffic lights, cars, and intersections. So we persevered and wished that the wind would at least blow from the side instead of head-on.

That darned crosswind

This was the silliest wish we could have had. A crosswind can be at least as annoying as a headwind. And it's insidious! You would think wind blowing from the side would simply be annoying without making it more difficult to ride. You might need to lean into the wind a bit more, no big deal. Unfortunately, that's not how it is! The reason is the tricky rule described above—namely, that resistance is squared in relation to the flow speed (in other words, when my speed doubles, resistance is fourfold). The speed of crosswind is also added to this equation, which means that we really are subjected to greater forces and have to pedal harder. (If you want to understand the equation in detail, we've provided a full breakdown in the sidebar at the end of this chapter. With the tongue-in-cheek heading "For Smarty Pants," these sidebars pop up throughout the book.)

Naturally, we weren't ready to just roll over and accept things as they are—especially since we have a true expert in our family. Sebastian Weber (Judith's brother, Marcus's brother-in-law) has trained triathletes and cyclists for competitions such as the Tour de France and IRONMAN in Hawaii and has developed his own digitized method for performance diagnostics and training planning.[2] He told us about special rims that change airflow and ensure that crosswinds actually help propel the rider forward. This only occurs when there is a very specific attack angle and minimal force. After all, the bicycle is only responsible for a small proportion of air resistance. The cyclists themselves are a much bigger factor.

US racing cyclist Greg LeMond experienced the most spectacular success with an aerodynamically refitted bicycle: In 1989, he won the Tour de France thanks to special aerodynamic handlebars. At that time, no one used handlebars that allowed riders to more or less lie on them, or helmets terminating in an aerodynamic point at the back. Throughout the tour, LeMond was in a head-to-head race with France's Laurent Fignon. One would take the lead, then the other. The distance between them was never more than a minute. Then came the final stage, the individual races on the Champs-Élysées. LeMond had a fifty-second deficit when he mounted a triathlon handlebar on his bike. It allowed him to lean far forward—an aerodynamically advantageous position. He also wore a teardrop-shaped

---

2   INSCYD, accessed September 10, 2024, https://inscyd.com.

helmet, the Giro Aerohead. Not only did LeMond make up the deficit, he even won a fifty-eight-second advantage. His victory was the closest in the history of the tour.

Anyone riding down a hill can experience the degree to which their sitting position impacts flow resistance. The smartest racing cyclists lean on the top tube between the front wheel and saddle, and position their neck close to the handlebar, in effect forming a human bundle that offers little flow resistance and also looks quite uncomfortable.[3] Some of these riding positions have since been banned because they pose too high a risk of injury.

If you wish to race downhill at even greater speeds and leave hard-riding competitors behind without exerting much effort, you need to place your hips on the saddle and stretch out your body in a perfectly horizontal, forward-facing position. This position results in an even greater reduction of flow resistance; the authors of the study cited above dubbed it the "Superman."

---

[3] Bert Blocken, Thijs van Druenen, Yasin Toparlar, and Thomas Andrianne, "Aerodynamic Analysis of Different Cyclist Hill Descent Positions," *Journal of Wind Engineering & Industrial Aerodynamics* 181 (2018): 27–45, https://doi.org/10.1016/j.jweia.2018.08.010.

But the "Superman" isn't allowed in competition, nor is it suitable for a cycling tour. Despite all our research, calculation, and cycling efforts, we were therefore unable to fully conquer the crosswind. Still, we were able to score a victory in some individual stages along the way, and we also gained more knowledge. Here are our tips for those who hate crosswinds:

- Close-fitting clothing can reduce air friction. And your bottom will be less sore in padded cycling shorts.

- Try to distract yourself from the laborious struggle against the wind. Instead, calculate how flow resistance changes when you cycle 2 km/h (1.2 mph) faster or slower.

- Be clever in how you plan your tours: Whenever possible, navigate the uncomfortable sections (guaranteed crosswind or uphill) first—they will be even harder to tackle when you're exhausted at the end of your tour. Plus, arriving with the wind at your back feels great!

- Embrace the feeling of heroism: Not only have you accepted that physics works against you when cycling, you've also understood why. And you're still cycling despite it all.

IMPACT SCALE

Annoying					
Lifehack					
Catastrophic					

• FOR SMARTY PANTS •

Crosswinds Can Come From the Front!

Imagine you're riding along at 20 km/h (12 mph). If there's no wind, the flow resistance generated by the airflow is 10 N. Now let's add wind at 20 km/h directly from the side.

How do airflow and crosswind add up to relative wind?

When wind blows from the side, airflow and crosswind don't move in the same direction; they cannot be simply added together to calculate the relative wind. We have to go back to an equation that will likely be remembered forever: $a^2 + b^2 = c^2$, also known as the Pythagorean theorem. We can use this equation because airflow (a) and crosswind (b) are perpendicular to each other. When we apply this, we get 400 (km/h)² + 400 (km/h)² = 800 (km/h)² [or 144 (mph)² + 144 (mph)² = 288 (mph)²]. The square root is roughly 28 km/h (17 mph), and that

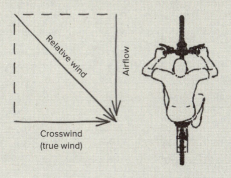

is the relative wind—a little stronger than the airflow but also a little weaker than if true wind were to come from the front. In other words, the wind factor calculated this way is roughly 1.4 times stronger.

What kind of force does wind now exert on us?

Wind resistance is calculated with the following equation: $F = \frac{1}{2} C_w A \rho v_r^2$.

C_w is the drag coefficient familiar to car enthusiasts, an expression of the streamlined shape of the vehicle. A is the area that comes into contact with the wind or "area of air impact," here the area of the bicycle including the rider as seen from the front, to be precise from the direction of the wind. ρ is the air density. This also has to be taken into consideration because in areas where the air is less dense, such as at higher elevations in the mountains, there will also be less wind resistance. Finally, v_r is the velocity (or speed) of relative wind. Importantly, and annoyingly, this figure is squared. Therefore, if we assume the following values ($C_w = 1.2$, $A = 0.55$ m², $\rho = 1.2$ kg/m³, and $v_r = 28$ km/h $= 7.8$ m/s), we arrive at air resistance of 20 N.

So how hard do I need to pedal?

The wind resistance we have calculated acts in the direction of relative wind—that is, in a diagonal backward direction. To understand the degree to which it slows down our forward motion, we need to separate the force into a component that acts laterally and a component that acts backward. Exactly the opposite of the addition of the two winds above. When we do this, we end up with the wind resistance in the direction of movement, which is $D_{forward} = D / 1.4 = 14$ N. The 1.4 results from the fact that we had to divide the forces in the same triangle as the winds calculated above.

The bottom line of this calculation is that although the crosswind itself doesn't come from the front at all, it does increase the flow resistance against the direction of movement from 10 N to 14 N. That sucks!

Toast, Tires, and Space

How one man tried to prevent the *Challenger* explosion— and why elastomers like it warm

Imagine you're a police officer standing at an intersection. A car stops and the driver asks you for directions: Should they turn right or left? You're familiar with the area and know that the correct route is to the right. Turning left leads to a precipice, the car will crash, and the driver will die. Of course, you point the driver to the right. You might even pull out your cell phone and show the driver that turning left will lead to catastrophe. The driver is annoyed, berates you—and then turns left. All you can do is watch helplessly as the car speeds toward the cliff.

Roger Boisjoly was an engineer, not a police officer, but he experienced something very similar that changed his life forever. The story begins in 1985 with a few sealing rings (or O-rings) that Roger Boisjoly was to examine on behalf of his employer. That employer, the company Morton Thiokol, constructed solid fuel rockets for NASA, among other projects. These rockets, also called boosters, were filled with a mix of substances (ammonium perchlorate, aluminum, and iron oxide to name but a few). They delivered most of the thrust to transport NASA's space shuttles into space. Once the fuel was spent, the boosters would separate from the shuttle, the shuttle would continue its flight, and the booster rockets would plunge into the ocean, from which they were then recovered and examined.

Roger Boisjoly had driven to Florida to perform one of these tests. Now he was holding the O-rings from the boosters for the *Discovery* shuttle in his hands and "almost had a cardiac arrest," as he would later tell *The Guardian*.[4] For the area around the O-rings was no longer bright, honey-colored, and elastic, but dark ("jet black" in the *Guardian* article), discolored, and pitted, as if someone had taken bites out of them. As an experienced engineer, he knew what this signified. Hot gas had been at work here— to be more precise, the gas the O-rings were supposed to have prevented from escaping. The seals were so damaged that Boisjoly wondered why the *Discovery* hadn't crashed.

You may ask yourself why a space shuttle needs rubber seals; after all, it isn't a canning jar. In truth, a canning jar and a spaceship have a few similarities. In both cases, sealing rings (or gaskets) ensure that the parts remain closely connected. When you close the lid of a canning jar

---

4   Mark Hayhurst, "I Knew What Was About to Happen," *The Guardian*, January 23, 2001, https://www.theguardian.com/science/2001/jan/23/spaceexploration.g2.

20 • THE UPS AND DOWNS OF PHYSICS

without a sealing ring (we're referring here to those nostalgic glass jars, like Mason jars, where the lids are fitted with a flat, wide rubber ring and attached to the jar with a metal clasp), it will always rattle because the lid and jar are both rigid, and small gaps will remain open between the two.

The same happens with a booster rocket, which is manufactured in four parts. Sets of two are preassembled by the manufacturer; the technicians at NASA then assemble the two halves on site and join them together. The sealing rings are placed between these two halves and ensure that no gaps form, even in the case of a slight deformation of the solid assembly halves. They are shaped like an "O" and run around the circumference of the rocket. This is why they are usually referred to as O-rings. The joint of the rocket halves is sealed with two O-rings, one above the other, or— as was the case with the *Discovery*—not sealed after all.

Roger Boisjoly immediately informed NASA as well as his own employer, Morton Thiokol, of his discovery. Then he set about exploring what might have caused the damage to the seals. Could the rings have been twisted? Unlikely— in a test they immediately bounced back on their own.

What could it have been? Boisjoly and his colleagues set up a fairly simple experiment: They placed one sealing ring between metal plates and squashed it slightly. Then they released the pressure and observed whether the ring retained contact with the two plates. As long as the temperature remained warm (the engineers experimented at a temperature of 100°F, or 37.8°C), the sealing rings performed effortlessly. But the cooler the temperature of the rings, the slower they expanded. At 75°F (23.9°C), they already needed 2.4 seconds to reestablish full contact, or a full seal. That is an unimaginably long time for their particular purpose. Much, much too long. Even a fifth of a second without contact would cause problems.

In the end, the engineers ran the experiment at 50°F, or a mere 10°C: "We stopped measuring after ten minutes," Roger Boisjoly recalls.[5] Together with his colleagues, he had found the problem. The *Discovery* had been launched at an exterior temperature of 52.88°F (11.6°C). The rubber rings were stiff from the cold. The first ring had failed to expand and hot gas had surged past, causing a "blow-by" in the vernacular of space travel. Luckily, the second ring had stopped the gas and averted disaster.

Why summer tires stiffen up in frosty temperatures

You can easily replicate the engineers' experiment at home by stretching two household rubber bands across a ruler and placing them in the freezer. Once you have taken them out again and removed them from the ruler, the rubber bands take quite a while before they retract to their original size. For these bands are elastomers, the term for synthetic materials with elastic properties. Like all synthetics, they consist of intertwined molecular chains.

---

5   This and the following quotes in this chapter are taken from a presentation given by Roger Boisjoly at the University of Minnesota in 1991 about the causes of the *Challenger* disaster. Source: "Unethical Decisions—The Causes of the Space Shuttle Challenger Disaster," posted April 12, 2020, YouTube, 1:21:08, https://www.youtube.com/watch?v=DB8iYf_857U. He begins speaking at the five-minute mark. Timestamp for this quote is 25:59. Timestamps for subsequent quotes from this source are provided throughout this chapter.

Imagine a plate of cooked spaghetti. Unlike spaghetti, the molecular chains in the elastomer are connected at many different places. This is deliberate because it is the only way to ensure that the rubber can expand and then spring back to its original shape. To achieve this, natural rubber is treated with sulfur, for example, to form bridges between the long molecules.

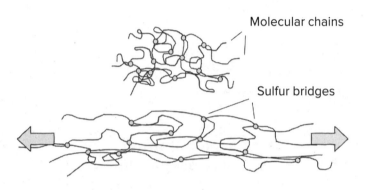

The rubber band will quickly snap back into its original shape, achieving the intended objective. But when elastomers are deformed and need to return to their original form, the molecular chains must be flexible or elastic. This elasticity is dependent on the temperature. If it's very cold, the long-chain molecules are less flexible—and take longer to spring back into their original form.

Elastomers have yet another property that complicates matters. When they are deformed (for example, when two parts of a rocket are pressed together), they discharge heat. But when they relax again, they have to absorb the heat. If the surroundings are cold, there is insufficient heat available for this process. The relaxing process is greatly slowed down, perhaps even halted altogether.

Elastomers respond differently to temperature variations, depending on the composition of the elastomer. This means summer tires for our cars turn rather stiff in cold temperatures and may have insufficient adhesion or grip with the road surface. Winter tires are manufactured from a different natural rubber composite and remain flexible even when temperatures are low.

To be suitable for space, a synthetic must be able to do two things: It must be able to sustain very high heat and it must expand rapidly. For the O-rings sit between two parts of the solid fuel rocket. They ensure that no cracks form through which the hot combustion gas could escape. In other words, contact between the O-ring and the parts of the rocket must be maintained at all times! Therefore, the synthetic used in the rocket O-rings is an elastomer based on fluorine caoutchouc, or FKM. It consists of long carbon chains to which the fluorine adheres. Fluorine creates very stable bonds, which is why it is especially stable when exposed to heat. But these elastomers are not calibrated for cold. In cold temperatures they quickly lose their elasticity.

We might be tempted to think that a danger spotted is a danger averted. Space rockets with these O-rings simply have to launch when temperatures are higher, which shouldn't be a problem in Florida. Boisjoly informed his superiors at Morton Thiokol and assumed that had taken care of the issue. Yet the next NASA mission was already planned: Half a year later, the *Challenger* space shuttle was to launch and set up communications satellites in space. In January—as you might imagine, not exactly the warmest month.

Just as Roger Boisjoly wanted to prevent the launch, others were determined that it should go ahead.

24 · THE UPS AND DOWNS OF PHYSICS

Boisjoly's first inkling was that very little action was taken following his report. If you watch any of the lectures he would later give at various universities, it is evident how the event weighed on his mind for many years after. Boisjoly is a large, heavy-set man, armed with numbers and data. An engineer who based all his decisions on facts and assumed responsibility. After all, many lives and a lot of money are at stake in a space mission. The planned *Challenger* launch was approaching, and no one had reacted to his data! Boisjoly could barely believe it.

At the end of July 1985, he wrote a memo to the Thiokol management in which he predicted a "disaster." In his talk at the University of Minnesota, he quoted from this memo: "'It is my honest and very real fear that if we do not take immediate action ... we stand in jeopardy of losing a flight, along with all the launch pad facilities.' ... I also mentioned in that memo that we were going to experience a failure of the highest order—loss of human life."[6]

Finally, action was taken and a task force was set up, though it only consisted of five engineers. According to Boisjoly, there was no support from the management. Nor did the results change, as was to be expected: The O-rings were too stiff to expand when temperatures were too low.

The *Challenger* space shuttle was to launch on January 28, 1986. Low temperatures—below freezing—were predicted for the Kennedy Space Center for that day. Freezing temperatures! NASA had never given the go-ahead for a launch at such low temperatures. Roger Boisjoly and his engineer

---

6   Timestamp: 14:19.

colleagues knew their data. They also knew what was to come: a gigantic explosion on the launch pad.

Squeezed between two tightly placed parts, the O-rings had to be able to expand in any and all circumstances, should small cracks appear between the parts due to the extremely high pressure inside the rockets or any other unpredicted vibration or shock. When temperatures drop below freezing, the elastomer is simply no longer an elastomer and cannot close the gaps. Due to the low temperature, the energy required to spring back into shape is simply lacking. There was another problem: Synthetics can only deform *in general* when the molecular chains are warm enough. Only then can they move a tiny fraction in one direction or another and not remain frozen in one position. If you cool a synthetic,[7] you reach the glass transition temperature—the temperature at which it is no longer fluid and flexible but hard and brittle, like glass. Above this temperature the synthetic remains more or less soft; below it this is no longer the case. This is why the glass transition temperature is also referred to as the softening temperature.

We replicated this phenomenon with a piece of rubber hose. We cooled the hose to −320.8°F (−196°C) with liquid nitrogen and then we hit the hose with a hammer. The hose shattered spectacularly into a thousand pieces.

You probably don't have liquid nitrogen at home. Luckily, we stumbled upon a similar experiment that anyone can try with a loaf of sliced bread. We always keep bread in the freezer so that we have plenty on hand. In a household

---

7   This applies to synthetics that are not fully crystalline.

of six, a single loaf simply disappears too quickly during breakfast. And because we're too lazy to slip the sliced bread, which is already packaged in plastic, into an additional freezer bag, we simply put it in the freezer as is. This is no problem for the bread, but it is problematic for the packaging. It becomes brittle in freezing temperatures. So you can simply break it!

Our freezer is in the basement with our washing machine, and when you balance a laundry basket on your hip with one hand, you can just about use your free hand to pull a frozen loaf of bread out of the freezer. If you hold the package near the front end, say roughly where the first four slices are, it simply breaks apart. And the frozen slices all tumble out onto the basement floor.

We would have known this if only we had read the warning on a package of "Golden Toast," a popular German brand. Below a mouth-watering photo of a stack of freshly thawed slices of toast, there is the following statement: "Use a freezer bag or cling film before freezing because our packaging loses flexibility at low temperatures and may tear." Bad luck for ignoring the warning! "Golden Toast" packaging is made of polypropylene (PP), while freezer bags are made of polyethylene (PE). PP and PE are both composed of long molecular chains. But in polypropylene, these repeatedly form small pockets in which they are very ordered, or crystalline, in arrangement. This diminishes the flexibility of the material, and the glass transition temperature lies between 32°F and 50°F (0°C and 10°C). A freezer's temperature is well below that, so this packaging is not suitable for balancing a frozen loaf of bread on top of a laundry basket.

I fought like hell

Of course, the outcome of this bread experiment is laughable compared to Roger Boisjoly's fears that the O-rings of the *Challenger* would fail right on the launch pad and hot combustion gas would escape, causing the tanks of liquid fuel to explode. He and his colleagues warned once more against launching in such cold conditions.

It was one day to launch. Hasty telephone conferences were called between the engineers and managers at Thiokol and NASA (it was 1986, so no video conferencing!). NASA requested a presentation—at such short notice that Roger Boisjoly could only bring his handwritten notes. Still, he was confident that the data were sufficient to halt the launch. He deliberately left out temperature information for giving the go-ahead, Boisjoly explains, because in the meantime he had concerns for all temperatures other than truly warm weather. Even 52.88°F (11.6°C) had been too cold, and now they were in winter! "That material gets just like a brick when it freezes."[8]

In everyday situations we are more familiar with this issue when synthetics get too hot—when we're ironing, for example. Many clothes have synthetics mixed in, usually polyester or polyamide. They are soft, keep their shape, and dry quickly. That's why they are so suitable for athletic clothing. Polyester can withstand a fair amount of heat (the melting point lies between 455°F and 500°F, or 235°C and 260°C), but if you fall wearing polyester track pants in the gym and slide across the floor, you might have a burn hole

---

8   Timestamp: 23:38.

or two due to friction. Polyamide, on the other hand, is very sensitive to heat and even a hot wash cycle (at 140°F/60°C, for example) can cause problems.

For the *Challenger*, the problem was the cold. The conference call on the day before launch lasted for six hours. The engineers presented their arguments, delivered facts, answered questions. Roger Boisjoly had the impression that the project managers at Thiokol were convinced the launch was a no-go. But then the mood turned. One of the NASA program managers asked, "When do you want me to launch, Thiokol, next April?"[9]

The Thiokol managers asked for a break in the conference. They wanted to consult without NASA for five minutes. Roger Boisjoly recalls this as follows: "The first words spoken when the ... button was pushed on mute were spoken by a general manager. He said in a soft voice, 'We have to make a management decision.'"[10] According to Boisjoly, the managers spent half an hour gathering a list of points that would justify a launch. The most important argument on the list was that the engineers' data were not sufficiently convincing.

The engineers in the room were not included in the discussion. At some point, Roger Boisjoly remembers, he got up and walked over to the managers. He threw the photos of the burned O-rings from the *Discovery* on the table: "I was told by my colleagues after the meeting that I was literally screaming at the managers to look at the photographs." To no avail. Later, the "senior vice president

---

9 Timestamp: 31:30.
10 Timestamp: 33:05.

TOAST, TIRES, AND SPACE · 29

general manager" was quoted as having said, "Take off your engineering hat and put on your management hat."[11]

Thiokol didn't want to lose NASA, an important client, and NASA didn't want to delay the launch. It had already been delayed several times because of poor weather, another mission, and technical problems. And so NASA was only too happy to accept Thiokol's acquiescence. When the managers announced their new decision, not a single question was asked. The conference call was terminated within a matter of minutes. Will had triumphed over facts.

The next day, on January 28, 1986, seven women and men boarded the *Challenger* and clipped into their seats. It was a bright sunny day, but a mere 35.6°F (2°C). Icicles hung from the scaffolding of the launch pad. Christa McAuliffe, a teacher, was among the astronauts. She had competed in NASA's "Teacher in Space" program and had been chosen from 11,000 candidates. Her task was to give a two-hour lesson from space via satellite. The first civilian NASA had allowed to fly into space! The entire nation joined her in feverish anticipation. When the all-clear was given for the *Challenger* to launch, roughly 17 percent of all Americans were watching live on their televisions.

At first, Roger Boisjoly was not among them. He had decided not to watch the launch. He remained convinced that the *Challenger* wouldn't even make it off the launch pad, and would explode immediately because of the rigid O-rings. But one of his colleagues, with whom he was friends, had a daughter who had never witnessed a launch and when they asked him to join them, he agreed. As the

---

11 Timestamp: 35:10.

30 · THE UPS AND DOWNS OF PHYSICS

countdown ended, the *Challenger* lifted off the pad accompanied by the cheers of the spectators. "We've just dodged a bullet," Roger Boisjoly whispered in his friend's ear. They watched the clock, counting the seconds, and awaited the catastrophe. Second by second ticked by and nothing happened. When the shuttle was in the air for a minute, his friend said a silent prayer of thanks.[12]

Thirteen seconds later, it happened. At precisely seventy-three seconds after launch, when the *Challenger* had reached a height of fifteen kilometers (nine miles), Boisjoly and his colleagues observed that a booster rocket seemed to be separating. His first thought was that it was too soon; that was supposed to happen after 120 seconds. Then the image on the television screen turned into a flaming ball. A booster rocket tumbled down to Earth in a cloud of smoke and fire. Nothing was visible of the shuttle itself at that moment. Only film and sound recordings, for instance by NBC News, capture the dismay of the commentators. The NASA ground crew were also crushed. For seconds, there was silence from mission control. Then a simple statement: "The vehicle has exploded."[13]

But that was not the case. Boisjoly's exact fears had come to pass. One of the O-rings failed a few seconds after the launch and a leak formed on one side. Hot combustion gas escaped from the leak. At first it seemed as if the leak closed up again (possibly even with hot ash or slag).

---

12 Timestamp: 42:37.

13 "Archival: Space Shuttle Challenger Disaster," NBC Nightly News, January 28, 1986, posted January 28, 2019, YouTube, 2:40, https://www.youtube.com/watch?v=yibNEcn-4yQ.

Otherwise, the *Challenger* wouldn't have even left the launch pad. But to the misfortune of the astronauts, the "plug" of compressed ash didn't hold. It was likely dislodged or dissolved when the shuttle flew through a strong gust of wind. Hot gases escaped and hit the joint of the booster with the exterior tank filled with hydrogen. Liquid oxygen and hydrogen leaked out—and immediately expanded greatly. This is why the accident looked like an explosion. Later studies determined that other factors also played a role, but this was the main cause of the disaster.

The crew cabin did not explode. But it couldn't be controlled by the crew either. The power failed and the capsule hit the ocean with incredible force and sank. It would be March before it was found with all seven crew members inside.

By that time, Roger Boisjoly had lived through many excruciating weeks. He had been called upon to join an investigative team but didn't get the impression that the true cause of the catastrophe was to be revealed. When the presidential commission convened by President Reagan questioned the Thiokol engineers, as Boisjoly recalls in his lecture, they had been instructed to respond to questions as briefly as possible. He decided not to follow this instruction. Instead of simply answering with yes or no, he handed over his files to the commission, including the memos in which he had warned of a catastrophe.

Hero or traitor

For some, this act made Roger Boisjoly a hero. His files enabled the commission to discover the true reason for the

Challenger crash. In June 1986, the commission presented its findings in a report that offered harsh criticism of NASA and named the O-rings as the cause of the catastrophe. (There is a famous scene from this presentation: Richard Feynman, Nobel laureate in Physics, placed parts of an O-ring in a glass with ice water to demonstrate how hard and rigid it remains. An experiment is worth a thousand words!) The report also contains previously unpublished photographs showing that in a matter of seconds after launch, small plumes of smoke were escaping from the lower connection of the booster rocket. These plumes eventually became flames which set the fuel tank on fire.

NASA's shuttle program was temporarily halted. Roger Boisjoly received the Award for Scientific Freedom and Responsibility from the American Association for the Advancement of Science.

But Boisjoly also paid a price. In the eyes of his colleagues, Roger was a traitor. Someone who was dragging the good name of Morton Thiokol through the mud, thereby threatening job losses. The fact that he lived in a town in Utah where Morton Thiokol was the most important employer made the situation even worse. Although he didn't lose his job, he was blocked from any access to the space program. He suffered from headaches, insomnia, and depression; doctors eventually diagnosed post-traumatic stress disorder. While his colleagues and neighbors were condemning him for having said too much, he was berating himself for having done too little. In the end, he resigned from Morton Thiokol and became an independent consultant. In his public lectures and seminars, he would return again and again to the most important topic of his life: ethics in natural and engineering sciences. As he told students at countless universities, "You will have a clear conscience and a peaceful sleep at night for doing the right thing."[14]

We were impressed by two aspects of this story: that a small part such as a rubber O-ring could have such a profound impact, and that it wasn't the laws of physics that caused trouble for NASA and the engineers—it was the fact that they ignored those laws! There is no negotiating the temperature at which a synthetic expands or stiffens—and the same is true for gravity, electrostatic charge, the greenhouse effect, and other physics phenomena. It's best if we all accept this fact.

IMPACT SCALE

Annoying		
Lifehack		
Catastrophic		

---

14 Timestamp: 55:30.

Hello? Hello? Are You Still There?

The difficulty of eliminating cellular dead zones

A call on the Intercity Express (ICE) from Berlin to Cologne: "Hello? Hello? (followed by a quick glance to check the cell phone)... I'm just past Berlin, might be cut off in a minute... Hello?... I lost you for a second (cell firmly pressed against the ear)... Hello? (voice rising)... I'll call again at the next station, OK?"

Yikes! Traveling in the "talking allowed" car of an express train, here a German ICE, where not just one but many such conversations are taking place, can be exhausting—especially since we all instinctively speak louder whenever we're worried about losing the connection. It doesn't help, of course, it's just a reflex. Why, we may well ask, is it so difficult to provide comprehensive, nationwide cellular reception? Why are there recurring dead zones every few kilometers in some regions and what could be done about it?

Our interest in this topic was piqued during a seminar at a conference center in Lower Bavaria. While we've forgotten the topic of the seminar, we remember what happened during breaks very well. The attendees jumped out of their seats and walked around in circles, arms aloft, running to the left and to the right. Until one of them called out: "Here!" Then everyone ran toward them, crowding around like they were getting ready for a game of musical chairs. But none of them wanted a chair, they wanted something far more valuable: cell reception. The general mood became increasingly fractious after each of these rituals or evening walks up the hill where at least those with subscriptions to Deutsche Telekom (German Telecom) were lucky enough to get a signal. The seminar presenter took to drinking gin during every break; if he hadn't, he might have done little else but try to check his messages.

Anyone who hears this story from us always has their own to share. Germany seems to be dotted with dead zones, a Swiss cheese of telecom reception. A study by Open Signal, which analyzes users' experiences with cellular reception, placed Germany fifty out of one hundred countries, below Indonesia and Kyrgyzstan. Rural areas in particular suffer from poor coverage. But users have also reported a 4G dead zone in the middle of Berlin! There's a rumor that at least one government minister refuses to be connected to calls with foreign colleagues while traveling in his car because he's so embarrassed that the connection drops all the time.

36 · THE UPS AND DOWNS OF PHYSICS

It's unclear how widespread the problem really is because there is no scientific definition of a "dead zone." Is it enough for there to be no network connection in a single street? Or does it have to be an entire village or community? And what about areas where Vodafone customers have no reception, but Telekom customers do? One thing is certain: Nationwide cellular reception doesn't exactly work smoothly in Germany, never mind sufficient reception for streaming. After all, we don't just want to be able to speak without interruption, we also want to be able to join a video call—ideally while our self-driving car is connected wirelessly to another autonomous vehicle. Would that even be possible?

Let's take a look at what happens when we make a call. Every time you call someone, your cell sends out electromagnetic waves. These waves spread out, searching for the closest transmission mast or tower. The area around these towers is referred to as a "network cell" where signals from cell phones are received. When you travel on a train from Berlin to Cologne, your cell phone browses from one network cell to another. From there, the signals are transmitted via wireless or cables to your colleague in his office, to your children at home, or to a foreign minister (if you happen to be a politician).

Our children played a game to try to spot these towers. In their mind's eye they imagined the classic image of one or several thin antennae, or satellite dishes mounted on buildings. But mobile towers don't look like that. They are more like a thick metal pole to which a series of strange longish gray boxes are attached. This shape is deliberate because the antennae *must* be elongated. For a trick of physics occurs inside each of these boxes! Several identical transmitters are located one above the other inside each box. If only one transmitter were used, the signal would spread out evenly in all directions—that is, upward, to the center, and downward. Roughly similar to the glow of a light bulb. But you don't want the signal below and of course above, especially if you live in the house below the tower.

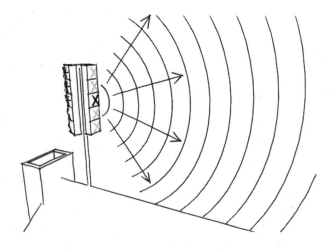

If we were engineers, we would be faced with the following question: How do we direct the signal so that it is transmitted in a forward direction as much as possible without being dispersed all over the place? In the case of lamps, the problem is solved with a simple lampshade, which allows light to shine through only in the desired direction. This is a little more difficult to achieve in the case of antennae. Still, there is a simple option to guide the transmission radius, one that is fascinating from the perspective of physics: Simply direct the waves of several transmitters at each other simultaneously so that they will keep each other in check.

This is why there are several antennae arranged one above the other on a single transmission tower. The radio waves overlap. If they are cleverly arranged, they strengthen in a lateral direction—that is, parallel to the ground—while wave accumulations (mountains) or valleys to the top or bottom are mutually weakened or even eliminated. This allows the service to be delivered to where it is needed, sideways from the sender (transmitter). It spreads parallel to the ground, allowing it to travel long distances—at least if there are no obstacles in its path, such as mountains, trees, or buildings. Generally speaking, we can even make calls while at sea up to thirty kilometers (nineteen miles) from the coast (at least in Germany).

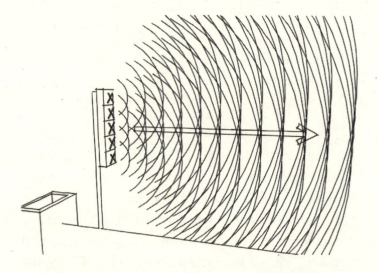

You may have discovered a mast that is installed at a great height (on a high-rise or on a television tower). In that case, we would want the signal to be directed downward. Of course, we could simply angle the box with the transmitters. But we don't even need to do that. All we need to do is harness the principle of superposition. When the waves from the lower sender are transmitted with a slight delay, the waves increase in intensity at an angle closer to the ground while at the same time canceling each other out where they aren't needed. Anyone who's been to a concert will be familiar with the long, narrow boxes mounted to the right and left of the stage. This superpositions the waves in an effective manner so that the good sound reaches the audience and not the ceiling of the venue.

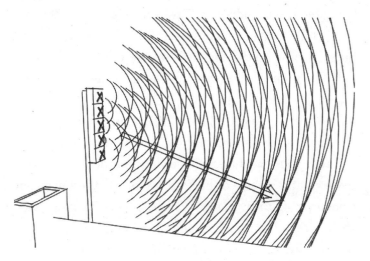

All this inspired us to locate our own personal antenna. Where might the closest tower to our apartment be? On our walks or whenever we went shopping, we would look up toward the sky, which must have seemed quite silly. We guessed that the tower would likely be to the west of our building because reception was always worst in our living room, which faces east.

We didn't find anything at first. So we entered the query "Where is the closest cell tower?" into a search engine. And voilà! The German Federal Network Agency[15] displays the exact location of the nearest tower (you can also consult www.cellmapper.net to find nearly any cell tower in the world). Here you can find your nearest antenna. In combination with Google Street View, you can easily discover where your own antenna is located.

Our very own tower

We entered our address and were thrilled to see that we had been right. The next tower is found to the west of our house. Unfortunately, a small, tree-crowned hill stands between it and us. To the east, on the other hand, there is no tower to be found for miles and miles. And when a tower finally appears, it's behind another house. That jibes with our day-to-day experience; we have hardly any cell reception on the east side (living room, bedrooms). If we want a stable connection, we must stand near our front door (on the west side). The best spot is next to a shoe shelf in the hallway, where the carpet is worn down from everyone hanging out to make calls. Of course, reception is fantastic in the attic, which provides a clear view beyond that pesky hill and the other house. (Our children's rooms are on the attic floor and it's very practical that we can easily reach them on their cell phones; their ubiquitous noise-canceling headphones prevent them from hearing

---

15 https://www.bundesnetzagentur.de/DE/Vportal/TK/Funktechnik/ EMF/start.html.

our old-fashioned analog call to the dinner table. A call on the cell phone works every time.)

But why does the tower on the west side manage to send signals our way despite the small forest, while the one on the east side fails to do so? Shouldn't it be that we get *no* reception on either side because the signal is blocked? We're grateful that this is not the case, but it doesn't seem logical.

We have physics to thank for the fact that the signal from the tower to the west manages to reach us. Radio waves have some tricks up their sleeves in much the same way as visible light. There's a lot of reflecting, dispersing, and deflecting going on. House walls can reflect the signal, the waves can be refracted downward by the edge of a roof, and uneven surfaces disperse your YouTube stream in all directions. This is why you never have good reception by one window in your living room, yet always have good reception at another window in the same room. The effect is most pronounced when a radio signal suddenly drops off at a traffic light. Sometimes, just driving a few meters forward resolves the issue.

Radio waves simply pass through some materials. The principle is that the thicker a wall, the more difficult it will be to penetrate it. Due to its built-in steel reinforcements, a concrete wall is far less permeable than a mortar wall. Conductive materials such as reinforced concrete are the stuff of nightmares for waves. Let's take the microwave. It "transmits" in a frequency range that is very close to Long Term Evolution (or LTE) and 5G—that is, precisely 2,455 GHz. The problem is evident: Do electromagnetic waves from your microwave spread unimpeded throughout your kitchen? No they don't, because they are blocked by the microwave's metal walls. We had a memorable experience of this phenomenon in our office. We'd put up a large whiteboard, and strangely enough, our Wi-Fi connection was very poor afterward. The link between these two factors was not evident to us—it was our IT administrator who discovered it. The wireless router was located on the other side of the wall on which we had mounted the whiteboard. The router was trying in vain to send its signal through the metallic whiteboard into our office.

Waves with superpowers

Why didn't the router succeed? It helps to take a quick look at how electromagnetic waves function. We can visualize them as many small electric and magnetic fields that are coupled. They oscillate perpendicular to the direction in which they spread. Waves of this type are emitted by your microwave, your cell phone, a flashlight, or a radio, and they differ in length. This has an impact on how effective they are in handling obstacles. Light and cell phone waves, for example, won't penetrate an aluminum suitcase, whereas X-ray waves do. Each wavelength has its own superpower and is therefore suitable for different purposes.

- **Long waves** are the giants among waves with a wavelength of up to 10,000 meters (32,800 feet). They transmit the time signal for our radio-controlled watches and can easily span a thousand kilometers (620 miles) because they run along the ground, following the curvature of the Earth.

HELLO? HELLO? ARE YOU STILL THERE? • 43

- **Short waves** are so short that they are reflected by the ionosphere. This allows the waves to be carried around the world. When we were children, we were captivated by the spy program broadcast on short wave. Anyone with a simple radio could listen in. But only spies with the correct key could decode the program and understand it.

- **Ultrashort waves** are what all of us use when we listen to the radio. And they are the busiest of all! Civilian and military communications, air traffic control, marine communications, and satellites are all controlled via these frequencies.

Cell phone signals consist of two "types" of waves: decimeter waves and centimeter waves. **Decimeter waves** (as you may have guessed) are at least a decimeter in length, measuring ten centimeters (3.9 inches) to a maximum of one meter (3.2 feet). Their frequency lies between 300 megahertz and three gigahertz. There is a huge amount of activity happening in this range: Countless transmission and navigation services (as well as your wireless network) send and receive via these frequencies. Whenever we're on stage performing science shows in our role as "Physikanten," our headsets also function on this frequency. Several radar installations use these very short wavelengths, and your microwave does the same! The decimeter wave's superpower is its data density. Even when individual frequencies lie close to each other, they don't interfere with one another. This makes it possible to load the waves with an enormous amount of data. Their weakness is that these waves are interrupted by objects that conduct electricity in the same dimension as the length of the wave, like large leaves on trees.

Centimeter waves (length 1–10 cm or 0.39–3.9 inches) are turbocharged radio waves. They make the extremely high rate of data for 5G technology possible in the first place. Shipping radar systems and TV satellites operate on these waves. Their superpower is their even greater data density. The disadvantage of centimeter waves is that the atmosphere can cause interference at higher frequencies. Vapor and rain reduce the range—although this can be utilized for weather forecasting. (Raindrops reflect the waves, which is why rain radar works so well.)

As you can see, waves experience different difficulties in finding their receivers. Very strong and tall towers were therefore erected for television and radio signals (VHF or very high frequency range), allowing them to span great distances. This made it possible, for example, to receive television from the West in former East Germany. Albeit not everywhere. Some GDR citizens lived in the "valley of the clueless." This referred to the area around the town of Greifswald (way up in the northeast of the country) and the district of Dresden. The electromagnetic waves transmitted by the broadcasters from the West simply didn't reach these areas. Citizens had to make do with (censored) state TV—and the acronym ARD, which stands for the consortium of public broadcasters in Germany, was often ironically dubbed "Außer Rügen und Dresden" or "Except Rügen and Dresden." On a fascinating sidenote, according to a study in which Stasi files were analyzed, the people in the "valley of the clueless" were more dissatisfied with the political system of the GDR than those who were able to receive TV from the West. This seems counterintuitive, doesn't it? The author of the study concluded that television broadcasts from the West were not seen as a source of information, but as entertainment and diversion. Being able to escape via media entertainment seems to have meant that people found the political system less onerous.

Today the German wireless network transmits images, movies, and a massive amount of data used for self-driving cars, video conferences, and the Internet of Things. At the time of writing, the 5G network is therefore being expanded. This allows for the transmission of a great deal of data in a very targeted fashion. The disadvantage is that 5G waves, which are transmitted in the centimeter-wave frequency, have a relatively short range. Consequently, it is necessary to build many more transmission towers.

This would solve our New Year's Eve problem. Every year, just before the clock strikes midnight, we try to call our loved ones who live elsewhere in Germany. And every year we fail to get a signal. This can't be due to the tower; it stands at the same spot as it does the rest of the year. But the network's capacities are limited. The mobile carriers try to guess how many cell phones are likely to connect to a single network cell. In areas with high population density, there are many towers—and, conversely, few towers in rural regions. If lots of people want to wish their loved ones a Happy New Year at the same time in our suburb in the Ruhr region, the cells shut down due to overcrowding, as it were. As a result, your cell phone may simply be unable to find a transmission cell where it can establish reception. So much for making a call. It's just like the dead zones in the ICE between Berlin and Cologne.

More transmission towers are not a problem from a technical and physical viewpoint, but they may well be so politically and economically. Towers have an image problem. Everyone wants to be able to make calls, but few want to have the antenna on (or near) their house. Again and again, there are protests when towers are set to be erected near schools, kindergartens, or animal habitats.

It is true that electromagnetic waves are radiated and captured wherever there's a tower. But the physics may surprise you: The more towers we erect, the *less* radiation there is. Sounds odd, doesn't it? But that's how it is. The reason is that the towers are always operated with the least possible output. Frequently this is only fifty watts (our microwave has 650 watts). This output is generally sufficient to distribute the signals within a small cell. However, when the towers are far apart, the output must be higher in order to transmit the signals. And that causes more radiation.

The situation is similar with a cell phone in a dead zone. When I am in a spot with no or very poor reception, my cell phone emits more electromagnetic waves, not less, because it is desperately trying to connect. Ergo, the more transmitters there are, the less radiation we are exposed to, because the cell phones don't have to "work" as hard.

My ears are burning

Is it really a problem if we are hit by electromagnetic waves? Physically, here is what happens: When I hold my cell phone anywhere near my body, the spot that is closest to the phone heats up a bit. This is because a water molecule is polar—that is, positively charged on one side and negatively charged on the other. When electromagnetic waves of a compatible frequency pass by, the molecule is set into rotation and puts its surroundings into motion—the water is warmed. This is a desirable effect in a microwave. But will I be cooking my brain with my cell phone? That's easy—really easy—to calculate.

For simplicity's sake, let's assume that our body consists of water (which it largely does). To warm one kilo of water by one degree, we have to exert energy of approximately 4,000 joules (or watt seconds).[16] The highest value with which the devices impact our bodies is two watts per kilogram. It's easy to calculate that it would take 2,000 seconds, or over half an hour, to heat the affected tissue by a tiny degree. However, this is a theoretical value because our circulation ensures that the heat is distributed directly (across our body). Consequently, little remains of the heating effect.

Still, warm is warm. And there's no disputing that something takes place between the cell phone and our body. Many people are unnerved by this. The good news is that there's no indication that anything happens to our cells, our nerves, or our DNA aside from minimal warming. There are no antennae in our body for cell phone radiation—at least as far as we know. At present, there are no methodologically clean studies that demonstrate a causal link between mobile transmission waves and disease.

In contrast to the calculation above, the human ear

---

16 The heat or thermal capacity of water (amount of energy needed to heat one kilogram of water by one degree) is 4.183 kJ/(kg K).

nevertheless gets quite hot during long calls! But this isn't because the cell phone exudes so much heat; rather, it prevents the ear from dispersing heat. It's as if the ear were covered in a hot blanket.

It will likely take some time before all dead zones in Germany are covered. Until then, all we can do is make the best of the situation. Some hotels located in deep dead zones offer holidays without cell reception, dubbing them a "digital detox." And rents for apartments and houses where there is no reception are also significantly cheaper.

IMPACT SCALE

Annoying	🌧️ 🌧️ 🌧️ 🌧️ 🌧️
Lifehack	💡 💡 💡 💡 💡
Catastrophic	💣 💣 💣 💣 💣

• FOR SMARTY PANTS •

Five Fun Facts for 5G

TikTok videos, music streaming, and phone calls: Everything that is transmitted is a powerful digital stream of numbers. This can be described as a very rapid sequence of "1s" and "0s" that works with the help of a radio wave, a so-called carrier wave. Information is thus transmitted at a specific, defined frequency.

In itself, a bare transmission wave carries no information. In order to transport data via the wave, we must change or modulate the wave. There are several options, two of which play a role here.

1. Amplitude modulation: This changes the size—or amplitude—of the wave. Simply put, if a "1" is to be sent, the wave grows big and for a "0" the size of the wave is very small or zero. This is possible in small steps and offers countless options.

2. Phase modulation: In this case, the wave is postponed slightly (hence "phase"). Depending on the type of shift, "0" and "1" can also be coded accordingly.

Practically speaking, both techniques can be combined and even expanded. So it is possible to send not only a single bit (that is, a "1" or a "0") per phase but, in the case of 5G, even eight bits (that is, a number between 0 and 255).

The great thing about 5G is that many other tricks come into play.

1. **High frequencies:** The use of high frequencies of up to 40 GHz is planned for 5G. Waves with a high frequency are short. This means that many more wave peaks and valleys reach the receiver in a short period of time. In other words, the wave can have a faster frequency or phase and transport far more information.

2. **Data highways:** Imagine that transmission on a frequency corresponds to a data highway. Why not use several? 5G makes it possible to drive on up to sixteen highways simultaneously. The data stream can therefore be divided into sixteen frequencies and

reassembled again on your cell phone. This shortens the download time of the next episode of your favorite series to nearly one sixteenth!

3. **Highways with thousands of lanes:** Before the data are sent, they are distributed across thousands of frequencies that are very close to each other. You can visualize this like very narrow lanes on a highway. Very little is transported in each lane, but just as much arrives at the destination overall. The great advantage lies in the fact that this technique, called orthogonal frequency-division multiplexing or OFDM, is far less vulnerable to disruption. If there's a disruption in a single lane (a sub-frequency), the complete signal can still be assembled for the recipient from the other data. Incidentally, a similar technology is used for QR codes. If these codes are a bit smudged or someone has scribbled something across them, they are usually still legible because the black dots contain redundant information.

4. **Multiple highways:** Even more data can be transmitted when the cellular base station sends the signal from several antennae, arranged close together, all of them in the same frequency. Ideally, this doubles the data rate—for example, when there are two antennae at the sender and the receiver (MIMO technology or multiple input, multiple output).

5. **Data transmission with beamforming:** Planning for the 5G network includes massive MIMO antennae in locations with high user density. Behind this is, for example, a box containing a chessboard-sized, eight-by-eight array of the smallest antennae. This allows the direction of the signal to be controlled. Receivers can even be supplied individually with wave beams. Since the massive MIMO antenna for one receiver sends out different signals depending on where you are, it operates physically very much like a hologram, which also looks different from varying viewpoints to create a 3D effect. Fantastic!

As you can see, enormous effort is being put into transmitting ever more data. Unfortunately, this also brings some problems. Because of the lower range, more cellular base stations are required, which in turn will use more energy overall. Add to this the anticipated increase in data volume. Both will lead to a rise in energy consumption.

Naturally, the enormous computing power required for the 5G network can't be managed with a Commodore 64. In fact, there are only a few chip manufacturers who can build the required electronics. The selection of suppliers for these important infrastructure building blocks has therefore become a highly politicized topic. For example, in the United States there is little confidence in the safety standards of Chinese suppliers—after all, we can't exactly look inside those chips. Some time ago, we were

52 · THE UPS AND DOWNS OF PHYSICS

in a New York electronics store and asked them to install a SIM card in our son's cell phone, which was made in China. The salesperson raised his hands in objection: "We're not allowed to do that."

A Collapsing Bridge? That's So Boring!

Why oscillations can have dramatic consequences

What would you do if a bridge collapsed next to you? Run away? Or stay and film it with your phone? Surely you wouldn't simply continue walking, bored and uninterested. But that's exactly what people did when the Tacoma Narrows Bridge collapsed in Washington State on November 7, 1940. Flickering black-and-white films (someone must have pointed a camera at the scene after all) show the 853-meter-long (2,798-foot) suspension bridge whipping through the air like a jump rope being swung with extra force. All the while, well-dressed gentlemen in hats and coats go about their business with no visible expression on their faces. Some even come from the direction of the swaying bridge, casually strolling off the deck. When they pass by the camera, their faces are clearly visible. Do they show fear or surprise? Not at all.

Obviously, people in the area had become accustomed to the crazy bridge. The Tacoma Narrows Bridge swayed back and forth at even the slightest hint of a breeze. Every single time. This was because of its construction; it was extremely long and extremely narrow. In fact, it was the third-longest suspension bridge in the world (surpassed only by the Golden Gate Bridge and the George Washington Bridge). Despite this, its deck provided just enough space for two lanes of traffic and a narrow sidewalk. The bridge would buck at the slightest wind, and it was soon nicknamed "Galloping Gertie." This peculiarity attracted tourists, some of them visiting specifically to "ride the roller coaster." This is likely why the passersby were so relaxed on November 7, 1940, when a slight windstorm arose and the bridge began to sway.

It's worth watching the historic films available online of the Tacoma Bridge collapse. In jittery black and white, Galloping Gertie gallops faster and faster until it finally breaks apart in the middle. A car that stood on the bridge until the last moment plunges into the water—the driver had already escaped to safety. Luckily no one was hurt, except for the architect's pride. Leon Solomon Moisseiff had planned the design very much in keeping with the time, because narrow bridges were "in" then. Cologne's Rodenkirchen Bridge over the Rhine has a similar design. So, what happened? After all, the Tacoma Narrows Bridge wasn't the first narrow bridge in the world. Had the architect done shoddy work?

The answer is no. But physics had gotten in the way— and bad luck with the weather. At the bridge, the prevailing wind came from the side, or crosswise. The deck started to sway and then it began to twist. The wind then hit the deck in this twisted position and it absorbed more energy, which amplified the swaying motion—physicists call this self-excited oscillation. Back then, there was no way to test bridge models in a wind tunnel.

Such tests are now available, and the renowned architect Norman Foster (the mind behind the cupola on the Reichstag in Berlin) probably felt safe when he designed the Millennium Bridge in London, an elegant pedestrian bridge across the Thames. It was inaugurated on June 10, 2000, and soon had its own moniker: "Wobbly Bridge." When too many people (nearly two thousand) crossed the bridge at the same time, it began to sway from side to side in a lovely rhythm, one swing to the left and to the right per second—one hertz. Two days later the bridge had to be closed for safety reasons. Once again, this was a kind of self-excited oscillation. When the bridge began to sway slightly as a result of all the pedestrians, people unconsciously adapted their gait to the rhythm of the oscillation—also called synchronization. Eventually they started to move almost like roller skaters. And this in turn amplified the oscillation of the bridge. The videos of the event are astounding![17]

We observed all these bridge catastrophes with amused nonchalance. The bridges we cross in the Ruhr region are not as beautiful, but they are stable, are made of concrete,

---

17 A quick search on YouTube for "Millennium Bridge wobble" leads to several videos documenting the phenomenon.

and don't "wobble." We didn't think self-excited oscillation would ever affect us—until, that is, we were attacked by our washing machine.

Galloping washing machines

It happened when our daughter came down to the basement to fetch the towels from the machine. The washing machine was already on the march. It hobbled toward her while in the spin cycle, tapping out a quick, steady rhythm on the basement floor. Our daughter called for help; we dashed down to the basement and had the momentary, delusional idea that we would be able to stop the machine in its tracks. How silly. If you've never touched a washing machine that's hobbling toward you while on a spin cycle of 1,200 rotations per minute, let it be known that it feels like holding a heavy jackhammer. In the end, all we could do was jump aside and watch in horror as the washing machine pulverized the laundry basket and left a dent in the wall. Only then did it come to a stop and offered no resistance to being pushed back into its corner.

At first, we thought that the forward movement of the machine was caused by the fact that the feet had gone slightly off balance over the course of time and were now at uneven heights. That turned out to be wrong, as we couldn't tip the machine by so much as a millimeter. It fell to our appliance technician—a genius who had often delighted us with inexpensive repairs to appliances we'd written off as broken—to explain that the damper inside the machine had given up the ghost. Dampers siphon off energy from the oscillation, thereby preventing it from keeping on growing. It was this function that had failed and enabled the machine to march across our basement.

The singing tea strainer

It should be easy to perform clever experiments with a phenomenon that lays waste to bridges and sets washing machines marching across the floor! For our favorite oscillation experiment, you need only two things: a tea strainer made of metal (stainless steel with small perforations) and a faucet.

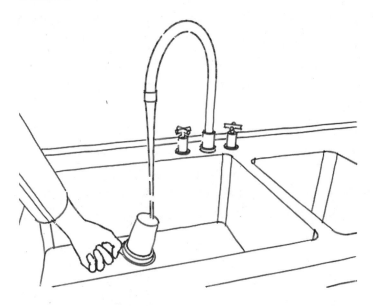

HERE'S HOW IT WORKS:

- Set the water running in your kitchen so that the stream is smooth. It shouldn't dribble, but there shouldn't be any bubbles emerging from the aerator (the piece at the tip that adds air to the stream to save water).

- Now take the strainer and hold it upside down at a slight angle, allowing the stream of water to hit the center of the surface. Adjust the height (of the strainer) and the strength of the water stream until you hear a singing tone.

This experiment would have been impossible ten years ago. Back then, loose-leaf tea was usually brewed in cotton nets or simply in teabags (as is still common). But since it has become fashionable to have teapots fitted with practical stainless-steel filters, tea drinkers are increasingly noticing that the strainers "howl" when all they want to do is wash out the remaining tea leaves. A television producer drew our attention to this phenomenon: "My tea strainer howls. Why is that?"

No question is more enticing to a household of physicists than "Why is that?" It's the best question in the world, an invitation to experiment, to ponder—and to explain. In the case of the singing tea strainer, our first experiments were frustrating. Our strainer didn't make a peep. Many failed attempts later, the methodology of chance came to our rescue. We had set the water stream exactly as described above and noticed that the strainer sang when we held it low in the sink.

What matters to the singing tea strainer is how fast the water flows. And it flows faster at the bottom because objects that fall accelerate as they fall. This is true both for a ball dropped from a balcony onto a patio and for water dropping from a faucet. For our tea strainer, the optimal water velocity is just above the bottom of the sink.

But where does the sound come from? The mechanism is the same as with a thin stick whipped quickly through the air. Our children love to do this when they get bored during a hike ("Do we always have to hike?"). The stick makes a satisfying "whoosh" sound!

This "whoosh" happens because the air streams around the stick at speed, resulting in many small, fixed vortices. When we say many, we really mean countless vortices, all less than a millimeter in diameter. And they rotate in opposite directions, one to the left and the other to the right. As soon as one vortex is released from the stick, another is formed. And the faster the air moves, the faster these vortices are produced. In the wake of the stick, a long sequence of counter-rotating vortices is created, a Kármán vortex street. This is always formed when air or water streams around an obstacle, and it is a beautiful sight. You can observe it in a fast-flowing creek, for example. Behind sticks or rocks peeking out from the surface, there are sometimes truly beautiful Kármán vortex streets.

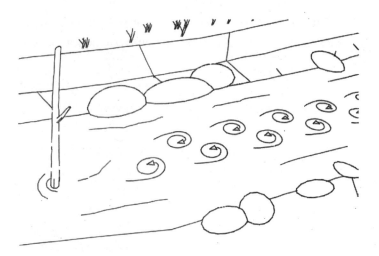

Where does the tea strainer sound come from?

The vortices in the creek don't make a sound (although the creek itself does make a lovely burbling sound). Why is it different with the tea strainer? To understand this, it helps to visualize the strainer not as metal with perforations but the reverse: a large hole with metal obstructions.

The water flows around the obstructions, just like the creek around the sticks. Vortices turning in the opposite direction emerge behind each obstruction, are then displaced by the water flow (a process called vortex shedding), and fall to the bottom. All the while, they continue to churn and turn, for vortices are incredibly stable! There is very little that can stop them; vortices experience hardly any friction with the water or air that surrounds them. Even simple, artificially created vortices in a pool can slide across the entire length (or width) of the pool,[18] and dangerous wake turbulence can also form behind airplanes, posing a danger to other planes, which may even crash.

Given all this vortex activity, it's easy to imagine that our stream of water flowing from the faucet is affected as well. All it wanted to do is flow downward in a lovely, even motion. The first "interference" came from gravity, which forced it to flow faster and faster. And then there are these

---

18 Pool vortices are a lot of fun! If you would like to try it for yourself, you can find a detailed description and fun stories about this phenomenon in our previous book *Physik ist, wenn's knallt* (published by Heyne, 2019) and on YouTube.

vortices! As they counter-rotate, the pressure inside the stream of water fluctuates greatly.

The metal connections in the tea strainer are shaken and rattled, and the entire piece begins to vibrate quite strongly. At first, the vibration may not be even, but after a little while, the strainer and vortices begin to sync. It's similar to what happened with the pedestrians on Norman Foster's Millennium Bridge in London, who adjusted their steps to the swaying of the bridge, which only amplified the oscillation.

For the tea strainer, this works as follows: When the water flows at optimal speed, vortex streets are formed, and the filter begins to sway in rhythm. This rhythm causes more vortices to form, which in turn amplify the oscillation, thus creating a stable system.

The key is that the tea strainer doesn't sway in a random fashion, but in sync with its natural or resonant frequency. This is the ideal frequency for a swaying object, its own inherent frequency. When you strike a bell, the sound it emits is always in the resonant frequency. When you tap a spoon against a tea mug, you hear its resonant frequency (or one such frequency, because if the mug has a handle, it will have several resonant frequencies. You can try this out by tapping different spots on the mug.) An egg cup has a different resonant frequency than a beer mug. And identical glasses filled with water to different levels have different resonant frequencies.

The bottom of our tea strainer vibrates in sync with its resonant frequency and emits the sound. Wilfried Suhr from the Institute of Physics Education at the University of Münster has published an excellent and comprehensive paper on the phenomenon[19] in which he proposes, among other ideas, that this experiment should be carried out in schools because of its simple, everyday nature. Unfortunately, this hasn't happened for our children yet, but there's always hope!

Music with vortices

The experiment deserves more attention because Kármán vortex streets and resonant frequencies can be used to make music. Tea strainers make different sounds depending on the shape of the bottom of the strainer, so they could be used for this. We tried to assemble a tea strainer orchestra—alas without success. After a shopping excursion through homeware stores in our town as well as various online shops, we were forced to conclude that the tea strainer industry isn't yet ripe for such an initiative.

But there is an instrument that functions according to the same principle: the wind harp or aeolian harp. The thickness and length of its strings are such that they vibrate when exposed to a certain wind velocity. All you need to do is to simply place them at an appropriate spot and music will play as soon as the wind blows. All by itself without anyone touching the instrument. King David (of the Old Testament) is said to have suspended an aeolian harp above his bed—although this raises the question of how windy his bedroom may have been.

---

19 Suhr, Wilfried, "Pfeiftöne vom Teefilter — Ein strömungsakustisches Alltagsphänomen," *Physik und Didaktik in Schule und Hochschule* 1/19 (2020): 57–66.

The people of Rotterdam experienced a similar, but less soothing, example of the phenomenon for themselves. The city is home to the Erasmus Bridge, which even at first glance resembles a harp: A 139-meter-high (456-foot), angulated pylon rises above the Nieuwe Maas river, supporting the deck with diagonal steel stay cables. Despite the thickness of the cables (up to 22.5 cm or 8.9 inches),[20] they began to vibrate just like the strings of the instrument. But not as a result of storms or the rhythm of pedestrians walking across; the vibration only started when the wind reached a strength of 6–7 on the Beaufort scale and it rained at the same time. One of the massive cables even swayed back and forth by up to seventy centimeters (twenty-eight inches)! And yet, as soon as the rain stopped, the vibration ceased.

What did rain have to do with the vibration? This was due to the water rivulets, which developed on the surface of the diagonal stay cables. They were small and shallow compared to the diameter of the cable. Still, it has long been known that even small surface changes can have a strong impact on aerodynamics. Golf balls, for example, fly twice as far as a smooth ball because of the dimples in their surface. When the cable began to vibrate, rivulets—one each along the top and bottom face of the cable—developed where air vortices are usually shed. A complicated system formed comprising the vibrating cable, rivulets moving

---

20 Chris Geurts, Ton Vrouwenvelder, Piet van Staalduinen, and Jaco Resusink, "Numerical Modelling of Rain-Wind-Induced Vibration: Erasmus Bridge, Rotterdam," *Structural Engineering International* 8/2 (1998): 129–135, https://doi.org/10.2749/101686698780489351.

along the cable as a result of the vibration, and periodic air vortices. The ideal prerequisite for such a system was to have a steel cable weighing a ton swaying wildly from side to side!

Luckily, physics offers a solution for every problem caused by physics: The Millennium Bridge in London and the Erasmus Bridge in Rotterdam were fitted with tuned hydraulic dampers, which prevented the vibrations from being amplified. The successful solution employed for billion-dollar structural engineering projects also worked for us at home. Thanks to new hydraulic dampers, our washing machine no longer moves so much as an inch.

IMPACT SCALE

Annoying	🌧 🌧 🌧 🌧 🌧
Lifehack	💡 💡 💡 💡 💡
Catastrophic	💣 💣 💣 💣 💣

Never Throw a Sofa out a Window

**Why we can't escape gravity
but can use it to mix a great cocktail**

Some stories seem to come straight from a cartoon, including an accident report submitted by one particular roofer. He had installed new roofing on a six-story building and seemed to have received more roofing tiles than he needed. Be that as it may, there were 250 kilograms (550 pounds) of roofing tiles left after he had completed his work. Understandably, he didn't feel like carrying all this weight down the stairs, so he decided to slowly lower them down the facade of the building. To this end, he looked for a large, stable barrel, attached it to a rope, and guided the rope through a pulley.

Then he ran down the six stories, attached the rope to the ground, ran back up, and loaded the barrel.

Once again, he descended to ground level—and untied the rope at the bottom. The bottom end of the rope was now

attached to the seventy-five-kilo (165-pound) roofer and the top end to the barrel with 250 kilograms of roofing tiles. It didn't remain at the top for very long. The barrel plunged downward, the roofer upward (still bravely grasping the rope). The accident report mentions that he encountered the barrel coming from above roughly at the third-floor level. When the barrel hit the ground, its bottom broke apart and the tiles fell through. Now the empty barrel no longer weighed 250 kilograms, perhaps not even twenty-five. The roofer at the top was suddenly heavier—and the whole process ran in reverse. He sped toward the ground while the remains of the barrel sped upward. Luckily, he survived with just a few broken bones. Still, what a story! And who was at fault? Gravity! It's the first answer that comes to mind for most people when asked what they know about physics. As we mentioned in our introduction, we have a postcard pinned to our office wall that reads: "Here's what I know about physics: Things fall and electricity hurts!" There is much truth in this. To us, it seems like a challenge to see whether it's possible to cheat gravity.

We decided to seek out a true heavyweight as an opponent (pardon the pun). Gravity is one of the four fundamental forces in physics. In contrast to the other three,[21] we experience it every day in our own bodies. Gravity defines our life—for good or ill. Gravitational force as the manifestation of the gravity at the Earth's surface is the reason why we fall. On a positive note, it also enables us to stay on the planet.

Children don't fall as far

Falling down isn't so bad in childhood. Small children tumble all the time and get up just as quickly. Their bones and joints are more flexible, and they have a smaller mass than grown-ups. And they fall from a lesser height. This means that the body has less time to accelerate while falling. The older we get, the less we like to fall. A reasonable attitude, because falling becomes increasingly dangerous as we age. According to online statistics, falls are the second leading cause of deaths in the home in the United States (31,400 in 2022). In fact, "the number of medically consulted injuries occurring in the home is greater than the total number of medically consulted injuries that occur in public places, the workplace, and motor-vehicle crashes combined."[22]

---

21 The other fundamental forces are electromagnetism, "weak interaction" (responsible for radioactive nuclear transformation, among other things), and "strong interaction" (also called "strong force"), which holds atomic nuclei together.

22 "Preventable Injuries and Injury-Related Deaths in Homes and Communities, 2022," Injury Facts, https://injuryfacts.nsc.org/home-and-community/deaths-in-the-home/introduction/.

Gravity can be a dangerous thing, even though it all began so harmlessly with Isaac Newton and his apple. (You're no doubt familiar with the apocryphal story that once upon a time in 1665 or 1666, Newton was lying beneath a tree and asked himself why an apple always falls straight to the ground. Some versions even contain the detail that this was a 'Flower of Kent' apple.) What is true is that Newton was the first to express gravitational force in an equation. Before Newton, there had been some pretty wild theories on what might compel bodies to fall. The Greeks, specifically Aristotle (384–322 BCE), perceived the cosmos to be a perfect, finite system, movements in general as a process that requires propulsion, and "matter" as a mixture of fire, water, earth, and air. According to this theory, heavy objects composed of earth and water fall to Earth. The heavier they are, the faster they fall. Conversely, according to this logic, the elements of air and fire strive upward.

Aristotle's world view went unchallenged for over 2,000 years because it fit very well with everyday experiences and provided a link between all areas (cosmos, movement, matter), seemingly without contradictions. Anyone who queried the details—such as Nicolaus Copernicus, who discovered that the Earth revolves around the sun as one of several planets—threatened to upend the entire construct, which was all the more compelling because it was upheld by Christianity. In the Middle Ages, Arabian scholars furthered the theory of a force between masses, although these ideas still hadn't found a foothold in the West.

Nutella doesn't fall faster

Enter Galileo Galilei. At the start of the seventeenth century, he ushered in physics as we understand it today. Galileo was among the first to carry out experiments in physics—for example, free fall. With his brilliant theoretical experiments, he questioned established doctrine. His most famous thought experiment was as follows: According to Aristotle, heavy objects fall more quickly than light objects. What would happen if we were to place a lighter object (like a chocolate bar) on top of a heavier object (like a jar of Nutella) and drop them? In theory, the chocolate bar should slow down the speed at which the jar of Nutella falls. On the other hand, the bar and the jar together (let's hope you keep this as a mere thought experiment in view of the precious content!) form a heavier body that should fall faster than the Nutella jar on its own. There is an obvious contradiction! And this led Galileo to realize that all objects fall at the same speed if you ignore air resistance.

But Newton clearly outperforms Galileo in the league of gravity researchers. In 1687, Newton published his first major work, *Philosophiae Naturalis Principia Mathematica*. The importance of this seminal work for physics cannot be overstated.[23] In it, he writes on the entire field

---

23 However, we don't want to overlook other important figures. Prior to Newton, the astronomer Johannes Kepler (1571–1630) had already developed three important laws that describe the path of planets. Robert Hooke (1635–1703) had the idea that the attraction between two masses is inversely proportionate to the square of the distance between them due to gravity. Both insights helped Newton in his thinking on gravitational force.

of mechanics and provides the tools for calculating almost any kind of movement. Students of physics spend nearly their entire first semester applying Newtonian mechanics and coming to admire the fact that all this is based on the famous three Newtonian axioms. But Newton's book also features the law of gravity for the first time, making it possible to mathematically express both a falling 'Flower of Kent' apple and the celestial bodies circling the sun. In other words, Newton's formula was in harmony with Galileo Galilei's ground-breaking observations.

Let's introduce the equation. Ladies and gentlemen, here is the first equation for any of the four fundamental forces (fanfare sounds!):

$$F_G = G \, \frac{m_1 m_2}{r^2}$$

If you'd like to know what the equation means and how to use it in calculations, check out the "Smarty Pants" section at the end of this chapter. For everyone else, all that matters for the moment is that gravitational force depends upon the masses of the two bodies involved and the distance between them. For gravity affects all bodies that have a mass, not just planets; it could be an elephant, your car, every single air molecule, or a barrel full of roofing tiles. And, of course, our own bodies.

Gravitational force at the kitchen table

As we wrote this chapter, we were sitting at our kitchen table, thinking it should be possible to calculate the gravitational force between the two of us. All we had to do was enter our respective weights and the distance between us into the equation! That's what we did and here are our results. When we sit across from each other at a distance of one meter (3.3 feet)—Marcus at a weight of eighty kilos (176 pounds) and Judith at sixty-five kilos (143 pounds)—the gravitational force between us is a whopping 0.0000003364 newtons. This corresponds to a force of thirty-four micrograms, a hilariously small force. It doesn't take much effort to overcome this gravitational force and push back the chair or get up and go to the kitchen to fetch a cup of coffee. At least we know we can easily conquer this gravitational force.

It feels a lot more challenging to get up in the morning and brew a pot of coffee. But in that scenario, we are up against the gravitational force of a cozy bed and, above all, the Earth. We can use the same equation to calculate how heavy the latter is. All we need to do is reverse the formula to arrive at the result: The Earth weighs 5.97 billion trillion tons. Sounds enormous! We clearly don't stand a chance against this weight. Still, although it's so much heavier than we are, we and the Earth are equal partners. The force acts upon both bodies. Just as the Earth exerts a force on us, so do we upon the Earth.

Darwin Awards and
their gravitational force

Unfortunately, this equality doesn't help us when we fall. Many spectacular stories about gravity end in fatalities. Take the lawyer who threw himself against a window on the twenty-fourth floor of a high-rise to demonstrate how safe the window was (it wasn't). Or the driver who wanted to quickly relieve himself by the side of the road when he was stuck in a traffic jam and jumped over a guardrail, but failed to notice the deep gorge on the other side. For more than thirty years, the Darwin Awards team has been collecting and presenting awards to the "stupidest" fatalities. The two examples above are taken from one of their lists. Not all instances are related to gravity. One criterion for the "award" is that a person died as a result of their own stupidity. But since gravity is ever present, it frequently plays a starring role.

From the perspective of physics, it's intriguing that we still don't know precisely how gravity works. We know that bodies attract each other—but how exactly do they do this? When it comes to other fundamental forces of physics, such as electromagnetic radiation, we know that they are based on the exchange of particles, so-called exchange particles that propagate forces from A to B. In the case of electromagnetic radiation, these are light particles or photons. Could it be similar with gravitational force? Scientists certainly surmise that this may be the case. They have already coined a name for the gravitation particles: gravitons. According to various theories, gravitons have no mass but an intrinsic angular momentum or spin. Thus far there is no proof of the graviton. But what is known is that gravitational force propagates at the speed of light, as Albert Einstein proposed with his theory of relativity. A few years ago, scientists succeeded in measuring gravitational waves for the first time. These are formed, for example, when two black holes merge, some one million light-years from Earth. This causes a distortion or bending of space-time. Albert Einstein would be amazed—he never thought it would be possible to take such precise measurements.

Why does food float on the ISS?

Despite this, we still haven't managed to outsmart gravity. Perhaps space will provide the key. After all, there's no gravity in space, is there? Yes, there definitely is! It's a widely held myth that there is no gravity in space. In its children's section, the *Stuttgarter Nachrichten* (Stuttgart news) website declared that "gravity doesn't work in space, that's why astronauts' food floats." The text aimed to explain why astronauts' sandwiches don't stay on their plates, but instead float through the interior of the International Space Station (ISS). All we can say is this: Children, don't believe a word! (Obviously this doesn't apply to all media, just in this particular case.) Gravity works throughout the entire universe. Everywhere. It must; otherwise, all the planets in our solar system would merrily drift apart.

It's true that the Earth's gravitational force diminishes the greater the distance. The ISS orbits at a height ranging between 370 and 460 kilometers (200 and 250 nautical miles).[24] It's understandable that the gravitational force is lower at that height than on Earth. The law of gravity helps us calculate how much less: 12 percent. In other words, at the ISS the Earth exerts 12 percent less gravitational force, but that still leaves 88 percent of the force that would

---

24 "International Space Station," NASA, updated May 23, 2023, https://www.nasa.gov/reference/international-space-station/.

be present if the space station were parked at your local supermarket.

Isn't this surprising? So why doesn't the space station fall? Because it's so fast. It moves at extreme speed—and if there were no gravitational force from the Earth, the ISS would simply speed straight ahead indefinitely. The astronauts are lucky that the Earth's gravity exists. It permanently draws the ISS to the Earth, slightly bending the orbit. At the correct speed, the movement of the ISS and the Earth's gravity cancel each other out and the space station moves around our planet in a beautiful circular orbit.

Or almost!

Gravity always wins and it does so in this case as well. This is why the ISS has to really rev the engine from time to time in order to avoid falling down to Earth. The flight height of the ISS over a longer period of time looks like this:

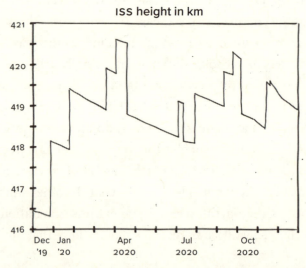

Where the curve soars upward in a vertical line, this indicates that the engine of the ISS was revved. For even at a height of 420 kilometers (227 nautical miles), the space station must contend with a certain amount of flow resistance. There is an atmosphere up there as well, even if it is much "thinner" than here on Earth. This atmospheric resistance creates friction—and it slows down the ISS. For this reason, the ISS has to accelerate sharply every few weeks to regain the appropriate height. Slowly but surely, it loses height, which is why the lines in the diagram slope slightly downward. The distance from Earth diminishes until the engine ignites again. Without this sporadic ignition, the equilibrium between gravity and centrifugal force would be disturbed and the ISS would orbit in ever smaller circles around the Earth, like in a vortex, until finally crashing down.

But we still want to know why the astronauts' food is floating! And not only that; the astronauts themselves are floating, doing somersaults in the air and buckling in when they want to sleep. They are weightless in the ISS. But that isn't because there's no gravity. On the contrary. For space stations and everything (and everyone) inside them, the same principles apply. Just like the shell of the ISS, the astronauts' sandwiches orbit the Earth at 7.66 kilometers (4.76 miles) per second and are subject to the same centrifugal force. Centrifugal force and gravity cancel each other out for both the ISS itself and the sandwiches. But in contrast to the ISS, the sandwiches remain relatively still: They float! If you nudge one of them slightly, it no longer has the same speed as the ISS, strictly speaking, and therefore grows either more distant or closer to the Earth. Yet this effect is so minimal that it isn't even noticeable.

A good thing too, because nutrition is very important for the astronauts. After all, they spend months in space and can't go shopping if they don't like what's on the menu. NASA operates a Space Food Systems Lab to develop meals that are non-perishable and tasty. The astronauts get to choose which meals they take on board. How satiating the food is depends once again on gravity. When we eat something, it usually travels to the pit of our stomach. Mechanoreceptors in the stomach then signal that something has "arrived." But when the food floats in the stomach that isn't always the case. The sense of feeling full takes a while to kick in; it only happens once the stomach has stretched slightly.

Gravity cocktail

At a friend's recent birthday party, a cocktail was served that can only be made thanks to gravity. (Yes, strictly speaking, gravitational force is the only reason that any beverage stays in its glass. But for this cocktail, it's especially true!) Here's the recipe:

INGREDIENTS:

- Orange juice

- Sparkling wine or water, tinted blue with food coloring

- Sparkling wine or water, tinted green with food coloring

- Pomegranate syrup (or any other red syrup)

INSTRUCTIONS:

Pour some orange juice into the glass. Carefully add the syrup with a spoon. The best technique is to hold the spoon against the inside wall of the glass just below the surface of the orange juice. Because the syrup is heavier than the juice thanks to its high sugar content, it sinks down in the glass. Now add the blue and green sparkling wine (or sparkling water), one after the other in the same fashion. These liquids have less sugar content than the juice and therefoi e remain at the surface. Now you have a beautiful, multi-colored cocktail.

The colors only blend when you stir the cocktail, which eliminates the differences in density. And if you stir vigorously, the carbon dioxide will even separate from the sparkling wine or water and the cocktail will no longer have that prickling effect on the tongue.

There is a similar phenomenon in nature, albeit deadly rather than amusing. In 1986, over 1,700 people died in Cameroon when large volumes of carbon dioxide were released from Lake Nyos. The lake is very deep, and its lower layers of water contain an extreme amount of carbon dioxide because they are subjected to a great deal of pressure from above. Just imagine a bottle of sparkling water with a tightly closed screw top. If the waters in the lake are "stirred up" for some reason, perhaps by a landslide or small earthquake, the water with a high carbon dioxide content rises to the top where the pressure is lower. The carbon dioxide can no longer remain dissolved and takes on a gaseous form. The gas bubbles now propel water to the surface and the entire lake is subjected to a momentous churn. In 1986, 1.6 million tons of CO_2 escaped into the air, flowed into two neighboring valleys, and killed many people and animals. Today, this region is a restricted area. To avoid keeping the restriction in place over the long term, long vertical pipes were installed in the lake. The water containing carbon dioxide escapes through the pipes into the air as a forty-meter-high (130-foot) geyser, continuously reducing the CO_2 content of the lake.

Escaping gravity

Can we humans outsmart gravity? Yes, we can! Here are three options:

- **Parabolic flight:** An aircraft lifts off with full thrust, then throttles the engines and describes a parabolic curve, first upward and then back toward Earth. During these first 25–30 seconds, the people in the plane are weightless. The pilot then fires up the engine again at full throttle. Everyone in the plane is pressed against the floor—until the next phase of weightlessness. This process is repeated some thirty times during a flight. Astronauts use the same principle to prepare themselves for weightlessness in space. Others pay thousands of dollars to experience what weightlessness feels like. All share in the experience of nausea, at least quite frequently. Before takeoff, they are given medication to settle their stomachs. Still, these flight simulators are flippantly referred to as "puke bombers."

- Anyone prone to nausea may prefer our next recommendation for experiencing weightlessness. Why not travel to the **center of the Earth**? In this model, you would be pulled by the same mass in all directions; to the left and to the right, up and down, and from the front and the back. The gravitational forces of all components of

Earth cancel each out. Unfortunately, you wouldn't fare well with the pressure, which would be higher by a factor of 3.6 million, nor with the temperature as high as 12,600°F (7,000°C). Maybe nausea is preferable after all.

- Insider tip: **Lagrange points**. Anyone who attempts to escape Earth's gravity needs to remember that the sun lurks just around the corner. At 300,000 times the weight of Earth, it's no small feat to pull free of the sun's gravitational force—unless you use the Earth's gravity to do so. For there are positions in space where the gravitational forces of Earth and the sun cancel each other out. The distances to the Earth and the sun remain constant at these positions, which we can calculate with great precision and even use in technology. They are called Lagrange points and we can count them on one hand, because there are precisely five.

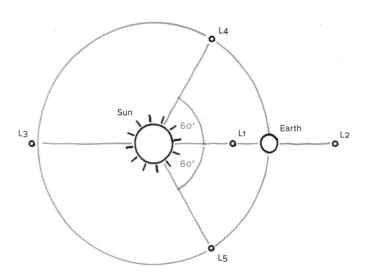

The successor to the Hubble Space Telescope, the James Webb Space Telescope, has been sent to point L2. The advantage is that it doesn't need to permanently orbit Earth like any run-of-the-mill satellite but remains on the other side, sheltered from the sun's radiation. Another Lagrange point, L3, has been inspiring science fiction authors since time immemorial. From the perspective of Earth, this point lies directly behind the sun and is featured in various books and movies as the possible site of a "Counter-Earth," a planet that looks exactly like Earth but which we cannot see because its orbit is always on the far side of the sun. This isn't plausible in terms of physics; after all, if there really were another Earth, it would also have a mass with its own force of gravitational attraction. This would disturb the equilibrium of the entire system. Nevertheless, the idea has given rise to several amusing stories. In the movie *Journey to the Far Side of the Sun* (1969), astronauts land on Counter-Earth and realize that everything is an exact mirror image of life on Earth. Furniture is on the opposite side of the room and organs are on the wrong side of the body. Even the German children's book *Urmel fliegt ins All* (Urmel flies to space) features a planet like this. It is called Arutuf, or "Futura" spelled backward. Could there be a reverse gravitational force on that planet (hypothetically speaking, of course)?

On our planet Earth, we can't avoid gravity. When we thought about the mishaps this causes, we remembered Henning Timmer, a singer in Marcus's school band. One day, decades ago, Henning was running through the streets of our town when a sofa fell from a window and struck him. If this conjures up a cartoonish animation in your mind's eye, don't worry. Fortunately, it didn't play out quite like that. The sofa only grazed him, and he escaped with a scare and scratches. He was lucky. For once a sofa is falling, it is firmly in the grip of gravity—and there's nothing we can do to stop it.

IMPACT SCALE

Annoying

Lifehack

Catastrophic

· **FOR SMARTY PANTS** ·

The Law of Gravity

$$F_G = G \, \frac{m_1 m_2}{r^2}$$

This is what the law of gravity looks like expressed as an equation. What does this law tell us? It describes the force of gravitational attraction that exists between two bodies as a result of their mass. In physics, forces are always given as "F." The abbreviation for **g**ravity is therefore $\mathbf{F_G}$.

This equation allows us to calculate the strength of the force of gravitational attraction between a person and the planet on which the person is standing. Let's begin at the beginning: On the right-hand side of the equation, we see the letter **G**—an extremely small constant painstakingly calculated in laboratories. Its value is 6.67430×10^{-11} m³ / (kg s²). The mass of each of the bodies in question—that is, the person and the planet—is more interesting for our purposes. In the equation they are represented as m_1 and m_2.

The greater the mass, the greater the force of gravitational attraction. The gravitational force is therefore proportional to both masses. If Earth were half its mass, our scales would only show half our weight when we check it as part of our morning routine.

The distance between the masses, here called r, must of course also be taken into consideration. The closer the bodies are to one another, the stronger the force of gravitational attraction. This is calculated from the center of gravity—in the case of a planet, from the center of Earth.

If the Earth's radius were half the size given the same mass, we would automatically be closer to its center of gravity. The gravitational force would increase dramatically! And that's not all. The equation contains r^2: At half the distance to the center of gravity, the gravitational force would quadruple. This is why conditions are so inhospitable near neutron stars with their incredibly high density.

Things would be slightly better if you were standing on Mars; the red planet has only one tenth the mass of Earth. This significantly reduces your own gravity on the ground on Mars. Because it's also only half the size of Earth, its center of gravity is much closer to you and the reduction of the force is somewhat negated. Taken together, these effects mean that scales on Mars would only show 0.38 the weight of what we weigh on Earth. One day, our children or grandchildren might be able to fly to Mars and try it out.

The Greenhouse Effect in Children's Rooms

Why windows let light in but don't let heat escape—and how this endangers our planet

When we moved into our new house, we were most looking forward to large windows and bright rooms. We would get to leave our dark attic apartment and enter a sun-filled row house with a yard. Okay, a small yard—but there are floor-to-ceiling windows in the living room with a French door leading directly onto a patio. But our children seemed to find it all *too* bright—their first act was to build a fort. And they did so in the living room, with no play taking place in the brand-new children's rooms on the top floor. Their grandparents had gifted them special cushions just for this purpose: rectangular pieces of foam, roughly one square meter in size and lovingly encased in home-sewn covers. With the help of these cushions, chairs, and many blankets, the children constructed a magnificent "house-within-a-house," emerging only to collect pretzel sticks and chocolate bars and drag them into their cave.

After three days conditions had deteriorated to such a degree that the cave dwellers were uncomfortable. "Someone should vacuum in there," our son declared, "but you're not allowed to dismantle the fort!"

The fort stood for another three days. Then our daughter accidentally sat on a partially melted chocolate bar and the children gave us permission to take it down. Together we gathered up the cushions. But when our daughter removed the last one from the window, she screamed. "The glass is broken!" It was true, there was a long crack in the windowpane reaching all the way down to the floor. We were mad as hell with the children—they must have hit the pane with a building block or something similar! But they swore that the hardest object they had taken into the cave had been the pretzel sticks.

Our anger now turned on the developer. After all, this was a brand-new house, and we still had a warranty, so to speak. We called, complained, and the pane was replaced. All was quiet for a little while. But a few weeks later the pane in the children's room had a crack. This room also had a large floor-to-ceiling window. This time we didn't even suspect the children but went straight to the builder. Why had they used such rubbish windows?

Yet this new complaint was rejected. "Did you lean something against the window?" the woman on the other end of the phone asked, clearly annoyed. "Perhaps a cushion or something?" Marcus was standing in the center of the children's room while making the call. Sure enough, one of the fort cushions was leaning against the window.

It was the same dark red as Marcus's face at that moment. Sheepishly, we ended the call and removed the cushion. The material was warm—it was early summer, and the sun was hitting the windowpane. It had heated the cushion and that, in turn, was intolerable for the window. But why did the pane tolerate the warm sunshine on the outside, when it couldn't deal with the warmth of the cushion on the inside? The answer to this question leads us deep into the science of physics. In the end, the incident with the cushion and windowpane even offered up an explanation as to why our planet is threatened by climate change.

Let's start at the beginning. The sunlight falls onto the window of the children's room and the window allows the light to pass through unimpeded (that is its purpose, after all). Still, this is the first moment at which we need to take a closer look. For there is no such thing as "the" sunlight. The light of the sun consists of an entire spectrum of wavelengths. Some are visible to us, others are not. In fact, we only see a shockingly small proportion of the light on the Earth's surface, as the graph below demonstrates.

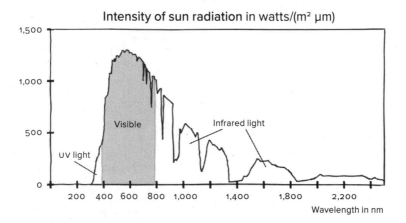

The x-axis represents the wavelength of light in nanometers (equal to one billionth of a meter). The y-axis shows the intensity of radiation with which this light hits the Earth. The shaded area identifies the spectrum that is visible to us! This visible light has a wavelength of roughly 380–780 nanometers. It contains a great deal of energy, greater than the light of all the other wavelengths combined.

On the bottom left, next to the visible light, the graph shows a small sliver that represents ultraviolet light. We can only see this light indirectly—for example, when it turns our skin red as a result of sunburn, or when it creates the glow of neon colors. To the right of the visible light, the graph shows the broad range of infrared light. These light frequencies do us no harm, nor are they visible to us. Some snake species are more capable than we are in this regard because they have a pit organ that enables them to sense infrared radiation. These creatures basically "see" a thermal image of their environment. Very useful if you're a nocturnal hunter. The pit organ is clearly visible on a snake's head: two small depressions to the left and right of the nostrils and the eyes.

Humans don't have anything like this, but we do have thermal imaging cameras. These are used to detect heat leaks in buildings or to take fun images showing that the tip of the nose is cooler than the forehead. If you like to tinker and would like to try this out, you can find instructions online to transform a simple computer webcam into an infrared imaging camera. This works because most webcams have a filter for infrared light. If you remove the

THE GREENHOUSE EFFECT IN CHILDREN'S ROOMS · 87

filter, you are left with an infrared imaging camera. You can also purchase an infrared imaging camera for your cell phone.

When we point an infrared imaging camera at that cushion by the window, we can clearly see what we had already felt: The material is very warm. This is because sunlight permeates the window in all its glory. Once it has done so, it hits the red cushion on the other side—unlike the pane, the cushion isn't light-permeable. It reflects the red light from the sunlight (this is why we see the material as red) and absorbs the rest. In other words, the cushion absorbs the light. The light radiation was halted by the cushion, but its energy remained. This energy now resides in the cushion, which turns hot.

Now the cushion has to do something with this heat. Compare this to a mug of hot chocolate that you're holding in your hands. There are several ways for the mug to disperse the heat, and it uses all of them.

1. The mug gladly passes warmth to your fingers via the simple mechanism of the fast-moving atoms in the wall of the mug transferring the energy of their movement to the skin of your fingers. This is **conduction**.

2. Warm air rises from your mug of hot chocolate. Cold air is pulled up along the sides of the mug. It touches the warm cocoa, is reheated by heat conduction, and once again rises. This heat distribution through air movement is called **convection**. Convection is very important—for example, in distributing the heat from your radiators throughout your home.

If these two options were the only means of heat transfer, we could construct the perfect insulation. We could pour the hot chocolate into a tightly sealed tin, suspend the tin from a very thin (and poorly conducting) string inside a vacuum chamber, and evacuate all the air from the chamber. Bingo!

Now the hot chocolate can't conduct the heat anywhere because it isn't touching anything (aside from the very thin string). Convection doesn't work either because of the acute absence of air. At this point, you've no doubt guessed that there's a catch. (You're right, the catch is at the top of the vacuum chamber. Just kidding!) There is indeed a third means of heat transport:

3. **Heat radiation.** Whether it intends to or not, every body emits electromagnetic radiation—in the case of the hot cocoa, this is infrared radiation. You could say this is infrared light, which we can't see. Hence

the cocoa inside the sealed tin keeps losing energy, or heat, until—well, until there's no heat left? Not really. Because *all* bodies emit heat radiation, the vacuum chamber itself and the room around it do the same. This means that the cocoa also absorbs heat radiation. The temperature of the hot cocoa only diminishes until it reaches the same temperature as its surroundings. At that point, heat loss and heat radiation cancel each other out, and all bodies in the space are in equilibrium.

Heat radiation is a complicated business. Until the end of the nineteenth century, experts tried in vain to come up with a formula to calculate how much and what kind of energy is emitted by bodies with differing temperatures. The famous physicist Max Planck finally succeeded in 1900 with his Planck's Law of Thermal Radiation, now known simply as Planck's Law.

Planck's discovery was revolutionary. At the time, many physicists were convinced that they had achieved a complete understanding of the world. When Max Planck embarked on his studies of physics, he was told that nearly everything had been researched and all that remained was to close some insignificant gaps. Nothing could have been further from the truth. Max Planck succeeded not only in covering all known contexts of heat radiation without contradictions, he also unwittingly discovered a new constant in nature: At the time, the speed of light ("c" in the equation below) and the Boltzmann constant "k" were already known. Under Max Planck, the Planck constant "h" appeared for the first time. This tiny number, which Planck inserted into his calculation, ensures that this equation makes it possible to calculate precisely how much energy (i.e., heat) a body emits. Incidentally, this was also the birth of quantum physics, although that would only become clear over time. The equation that expresses Planck's Law is as follows:

$$B_\nu(T) = \frac{2h\nu^3}{c^2}\frac{1}{e^{h\nu/kT}-1}$$

A bit intimidating, isn't it? Don't panic! As with most equations, the meaning is really quite simple: It states that a body with a specific temperature *always* emits energy in a specific wavelength. Max Planck was operating with an imagined "ideal blackbody." When we enter the temperature of the sun (approximately 10,830°F/6,000°C) into the equation, the result is a radiation spectrum that resembles the graph earlier in this chapter with visible and invisible (infrared) sunlight.

Our hot cushion is also a body that emits energy. Unfortunately, we couldn't measure its precise temperature because we only noticed the damage to the windowpane when it was too late. However, we could estimate it (and physicists love to estimate when they don't have the precise answer). Here we go: Our window is composed of standard double insulating glass. Panes like this can tolerate differing heat exposure at different points on the glass. After all, the sun shines on the upper left corner in the mornings and then travels down over the course of the day. Still, this tolerance has a limit.

According to the manufacturer, our pane has a thermal shock resistance of 72°F (40°C). In other words, the

THE GREENHOUSE EFFECT IN CHILDREN'S ROOMS · 91

temperature on the left side of the pane can be 68°F/20°C (= normal room temperature) but 140°F (60°C) on the bottom right, where the cushion was located, without causing a crack. Therefore, the difference must have been greater in our case because the pane did crack. Let's assume that the cushion increased the temperature of the pane to 158°F (70°C).

Now Max Planck comes back into play! When we enter a temperature of 158°F (70°C) into the calculation, we find that the cushion radiates the most heat at a wavelength of 8,500 nanometers (nm). This radiation is directed primarily at the window against which it is leaning. The cushion has "solved" its energy problem: temperature radiated, mission accomplished!

Unfortunately, now it's the window that has a problem because the cushion transfers the radiation in a wavelength with which the window cannot cope. At 8,500 nm, we are not in the realm of visible light, but only pure heat radiation. Our brand-new row house, flooded with natural light, is of course equipped with energy-efficient windows. These are designed to keep heat *inside*.

Here[25] you can see how light transmittance works in modern windowpanes. Visible sunlight from the outside with a wavelength of 380–780 nm is allowed to pass through to ensure that the interior is nice and bright. A large proportion of infrared light, invisible to us, also

---

25 M. Rubin, "Optical Properties of Soda Lime Silica Glasses," *Solar Energy Materials* 12, no. 4 (September–October 1985): 275–288, https://doi.org/10.1016/0165-1633(85)90052-8.

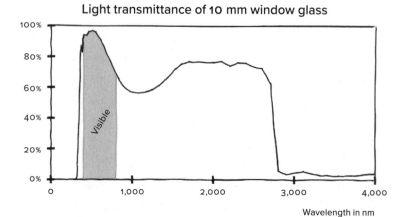

permeates the window (which is why you can turn on your TV with an infrared remote control even through a pane of glass). But pay attention to the right end of the y-axis of the graph, which marks 4,000 nm, a wavelength at which the pane basically allows *no* transmittance. The same is true for even greater wavelengths. The wavelength of 8,500 nm emitted by our cushion doesn't pass through the glass. Simply put, the window lets sunlight pass through but doesn't allow the heat to escape.[26]

Of course, the cushion is "unaware" of these factors. It continues to radiate heat. The energy builds up at the spot where the material touches the window, and the glass heats up more and more. And what does the pane do? It tries to expand, as warm bodies are wont to do. But the glass has little space to expand. Consequently, pressure keeps building until it finally cracks.

---

26 This is, of course, an ideal scenario for oven windows. They, too, are not supposed to allow heat to escape.

You can replicate this effect in a destructive experiment by quickly pouring boiling water into a glass (make sure to place the glass in the sink or in a bucket, and do not hold it in your hand). Provided you aren't using a latte glass or a mason jar with thick walls, chances are good that it will break. The glass becomes very, very hot at the spot where the hot water touches it. The glass tries to expand but can't do so quickly enough, which is why it cracks. However, if you pour the boiling water into the glass slowly, pausing every now and then, the glass will survive.

The pasta sauce debacle

We recently undertook this experiment, albeit unintentionally. We have an office tradition where the team has lunch together every Wednesday. Everyone takes turns to bring lunch or to cook at the office. Although we don't have a fully equipped kitchen, we do have a galley kitchen with a sink, fridge, and two electric hotplates. We wanted to heat up some pasta sauce on one of these hotplates (a delicious creamy sauce with mushrooms and cherry tomatoes). Unfortunately, the delicious sauce was in glassware, which is perfectly suited for baking, but the localized heat of our electric plate turned up high was clearly too much for it. A neat crack formed all around the pot at roughly one third of its height. The creamy sauce poured down the sides of the small fridge on which the electric hotplate sits. Luckily, we always have a jar of pesto in the fridge.

There are worse things than spilled pasta sauce! Sunlight pours in, the heat can't escape—does this remind you of something? Exactly, the greenhouse effect. The same phenomenon we observed on the window in our children's room is now happening on our planet, albeit with gas instead of glass. Various gases that we produce on Earth, and which gather in the atmosphere, play the role that the windowpane played in the other scenario. Carbon dioxide is one of them, as well as methane, nitrogen oxides, and water vapor. Although these gases represent less than 1 percent of the atmosphere, they prevent heat from escaping from the Earth into space. Instead, the heat is absorbed by the gaseous layer. It then radiates in all directions, including back toward the Earth, which is getting warmer and warmer.

Greenhouse Earth: Gas, not glass

One gas in particular is responsible for the greenhouse effect. Water vapor causes roughly two thirds of the natural greenhouse effect, and it is part of a vicious circle in which climate change heats the oceans and other bodies of water. The warmer they become, the more water evaporates—and the more water vapor reaches the atmosphere. But the atmosphere can only absorb a finite amount of water vapor. At the same time, the warmer the atmosphere becomes, the more water vapor it can store. In turn, the increased amount of water vapor blocks even more heat from escaping into space. Instead, it is reflected back to Earth. A positive feedback cycle, a self-reinforcing process that keeps on heating the Earth more and more, much like a greenhouse.

Gardeners know that simple greenhouses need to be sufficiently ventilated and equipped with blinds and coverings. Judith's parents recently learned what a greenhouse without such features can do. With the help of an architect, they installed a greenhouse on the flat roof of their garage. This wasn't a small, prefabricated greenhouse but a custom-designed, modern structure made of beams and glass that covered nearly the entire surface of the garage roof and was accessible from the top floor of the house. The greenhouse would allow large potted plants that grace the terrace in warm weather to thrive during the winter months. They had a lemon tree, an olive tree, and a large plumbago shrub with light blue blossoms (*Plumbago auriculata*, also known as Cape leadwort) —all plants that originate from warm climates and don't appreciate the German winter. The idea was that they would overwinter in comfortable temperatures and sufficient light.

There was indeed sufficient light.

However, as far as the temperatures went, the architect had clearly underestimated the greenhouse effect. The plants were scorched. Even in winter! All it took was a few rays of sun to transform the structure on the garage roof into an inhospitable death zone. The fact that the glass walls were angled and thereby increased the solar irradiation (which is dependent on the angle of incidence) only served to enhance the greenhouse effect.

After the first winter, Judith's parents discarded the scorched plants, bought new ones, and installed blinds on the greenhouse. For several years, it served its purpose. Unfortunately, another problem began to emerge over time: transporting the heavy pots up to the garage roof. These aren't flowerpots you can just carry up the stairs. The lemon tree, for example, stood in a clay pot holding

500 liters (132 gallons) of soil with an estimated total weight of 750 kilograms (1,650 pounds).

The architect had installed a pulley system. A pole with a pulley at the top was attached to the garage wall. A rope ran through the pulley, and this is what Judith's parents used to heave the heavy pots up to the greenhouse. Judith's father tied the pots to the rope at the bottom and pulled, and her mother stood on the roof and took receipt.

At some point, the pole must have pulled loose. At any rate, when Judith's mother wanted to untie a pot at the top and leaned onto the pole, the pole broke away from the wall and she fell off the roof. Luckily, a wheelbarrow filled with earth awaited her at the bottom, as well as Judith's strong father. She landed directly in the wheelbarrow and was unhurt. Still, ever since that day no more plant pots were pulled up to the garage roof.

The greenhouse was modified again when some of the glazed walls were replaced by solid walls. The house had always needed an extra space for a study or home office. The lemon tree now spends its winters in the garage under a grow lamp that simulates Mediterranean sunshine from 8 AM to 6 PM. The olive tree stays outside in winter and seems to be surviving thus far—climate change being in its favor in this case.

IMPACT SCALE

Annoying	🌧	🌧	🌧	🌧	🌧
Lifehack	💡	💡	💡	💡	💡
Catastrophic	💣	💣	💣	💣	💣

High-Rise Melts Car Parts

Why the burning glass effect can be downright dangerous, yet still helpful for applying makeup

Imagine the mouth-watering sound of eggs frying, the egg whites almost firm. Tasty, if only the frying pan hadn't been sitting on the hood of a car in the center of London. It was spring 2013 and the city's financial district was hot. Swelteringly hot! It wasn't climate change that was at fault, but a high-rise.

High-rises tend to be controversial because they block the light from their neighbors. But this one did the opposite: It radiated bundled sun rays down to the ground that caused bicycle seats to melt and enabled journalists to fry eggs on a car. It even melted the paint on a Jaguar, which led to several parking spots being blocked off.

The reason for these death rays, as the British press dubbed them, was the unique shape of the building at

98 · THE UPS AND DOWNS OF PHYSICS

20 Fenchurch Street; the south facade consists of mirrored glass and is also concave (that is, curving inward). In 2013, this was an innovative design because it had only recently become possible to construct concave facades. The cost of the project was accordingly high: 200 million pounds (then about US$313 million). The latest addition to London's built environment was a 160-meter-tall (525-foot) tower with offices, a restaurant, a sky garden, and a viewing platform.

But the locals weren't particularly grateful. Even during construction, the death rays had caused early accidents. Chairs outside cafés started to smolder, tiles shattered, and carpets were in danger of catching fire. Passersby recounted feeling like their outstretched hands were burning. Film footage shows thermometers measuring a temperature of 115.7°F (46.5°C) in front of the building in the shade. Clearly, just a few rays of sun were sufficient to turn the high-rise into a burning glass.

It makes sense from the perspective of physics. The south facade of the building is concave and made of mirrored glass. And this facade acts exactly like any concave mirror would, gathering as much light as possible and reflecting it to focus on a single spot. This spot is called the focal point or combustion point—a fitting name in the case of the London high-rise.

Had the facade been a sheer vertical, the effect would not have occurred. Light that falls onto a normal, straight mirror is reflected at the same angle as the angle of incidence (the angle at which the light hits a surface). The distribution of the reflected light is the same as the distribution of the incident light. This is why most high-rises don't "fry" their surroundings, even if their facades are clad in mirrored glass.

Things are different with a curved mirror. Imagine the mirror as an entire sphere. Now place a dot-shaped source of light at the center: The walls of the sphere now reflect the light right back to the center of the light source.

Of course, the curvature on the facade of the London high-rise isn't as pronounced as a sphere. But we can imagine the curved facade as a tiny section of a sphere. This section is sufficient to bundle the rays, and it is known as a hollow spherical mirror or concave mirror.

To understand what happens when a light bundle falls onto a concave mirror, we have to look at the rays of light individually. When a ray hits a concave mirror, the curvature has no effect at first. It is reflected according to the familiar principle of "angle of incidence equals angle of reflection." It only heats up when we also pay attention to the trajectory of the other rays. For they all overlap at a single point, the focal point. This is where the energy of many rays is bundled—and that can get very, very hot...

Shop owners in the streets adjacent to the high-rise installed scaffolding with black netting in front of their shops to protect them from the rays. The high-rise itself also had to be retrofitted; louvres were installed on the south facade to prevent the solar rays reflecting. The interior of the building, by contrast, was comfortably cool— after all, the star architect from Uruguay, Rafael Viñoly, had designed a facade with mirrored glass.

But Londoners weren't satisfied with these measures. In 2015, they awarded the high-rise the "Carbuncle Cup,"

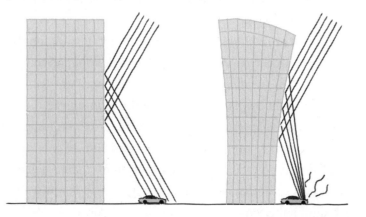

a prize given to Britain's ugliest building (as the word *carbuncle* vividly suggests). Aside from the death rays, there were several other criticisms:

- The building is surrounded by designated heritage sites in an area where the heights of new buildings are restricted.

- It is narrow at the bottom and wider at the top, which earned it the nickname "Walkie Talkie." *The Guardian* derided its shape as a "diagram of greed"[27] whose only purpose was to provide more interior space for the upper, more expensive floors.

- Passersby and neighbors complained of unusual whistling sounds, which seemed to emanate from the light shafts.

---

27 Oliver Wainwright, "Carbuncle Cup: Walkie Talkie Wins Prize for Worst Building of the Year," *The Guardian*, September 2, 2015, https://www.theguardian.com/artanddesign/architecture-design-blog/2015/sep/02/walkie-talkie-london-wins-carbuncle-cup-worst-building-of-year.

- The Walkie Talkie was blamed for unusually strong wind gusts, knocking over shop signage and patio furniture and even pedestrians. This could have been caused by "fall winds" from the tall building being channeled down to the ground.

It's interesting to note that London wasn't the only place where Rafael Viñoly had made the mistake with the concave facade. He had already designed a similar facade for a hotel in Las Vegas in 2010. In that instance, the concentrated rays were channeled into the pool in the courtyard, albeit not as a nifty natural heating source for the water. Instead, the rays hit pool loungers, melted plastic flip-flops, and chased tourists to seek shelter in the shade. So it's all the more astonishing that he chose the same concave form for the Walkie Talkie. As Sartre is believed to have said, "One should commit *no* stupidity *twice*, the variety of choices is, in the end, large enough."

To save the architect's honor (after all, he has designed many buildings without death rays), it's important to note that the burning glass effect occurs again and again, unintentionally and surprisingly. While we were researching this book, the British physicist Wendy Sadler sent us a photo of her makeup mirror, set up by a window to benefit from natural light. There is also a burn mark on the wooden window frame, clearly visible in the photo.

Makeup mirrors are also concave. They have a curved mirror surface which creates a magnifying effect. This was used by physicians over 300 years ago to look more closely into patients' nostrils and throats. In Wendy's case, the light bundled by the concave mirror was obviously too much for the wooden window frame. Wendy isn't alone: Berlin's *Tagesspiegel* newspaper dedicated an article to the magnifying glass effect after editor-in-chief Lorenz Maroldt's apartment was nearly set on fire, also a result of a mirror in the living room.

By the way, the term "burning glass effect" isn't accurate: A concave mirror is not a burning glass. A burning glass is transparent glass that is convex like a magnifying glass, which means it curves outward. Of course, both concave mirrors and convex lenses can be dangerous. You don't even require a perfect lens to cause a burning glass effect. In summer 2019, an apartment in Germany burned down because, it was assumed, some bottles had been stored on the balcony for too long. According to the police report,[28] the bottles just happened to be placed in such an unfortunate manner that the energy of the bundled light was channeled directly onto some cartons behind the balcony door inside the apartment. This is why the police warn the public not to leave bottles lying around in forests and fields during dry, hot summers.

But when we channel light intentionally at a curved glass, we can carry out a lovely experiment.

YOU WILL NEED:

- Reading glasses, ideally of the cheap drugstore kind

- A source of light, such as a desk lamp

---

28 Ingo Rodriguez, "Brennglaseffekt entfacht Feuer: Pfandflaschen verursachen Wohnungsbrand," *Hannoversche Allgemeine Zeitung,* June 23, 2020, https://www.haz.de/lokales/hannover/brennglaseffekt-entfacht-feuer-pfandflaschen-verursachen-wohnungsbrand-S36HHIBSR6U7DTAUWFHFKPX4FU.html.

HIGH-RISE MELTS CAR PARTS • 103

HERE'S HOW IT WORKS:

- Turn on the desk lamp and direct the light at a wall or door several meters/yards away.

- Hold the reading glasses so that the light shines directly through them and you can see a shadow of the glass frames on the wall.

- Move the reading glasses back and forth, closer to the wall and further away again. Note the point at which a focused shadow of the lamp appears on the wall.

THE SCIENCE BEHIND THE EXPERIMENT:

The distance at which you need to hold the reading glasses from the wall depends on the strength of the glasses, which is measured in diopters. A diopter is a unit of measurement that expresses the reciprocal value of focal length in meters. This means that the focal length is half ($1/2$) a meter for two diopters ($2/1$). In this case, parallel rays of light are brought to focus at exactly fifty centimeters (19.7 inches) behind the reading glasses. This works perfectly when the light source is at a great distance, like the sun. But when the light source is in your room, the sharp or focused image occurs slightly behind the focal point. Despite this lack of precision, this method is nevertheless useful to estimate the strength of reading glasses for people with hyperopia (farsightedness). Simply take the distance between the glasses and the focused image on the wall in meters and calculate the reciprocal value. If the distance is roughly one meter (3.28 feet), the glasses have one diopter.[29]

Concave mirrors have one advantage over lenses. While lenses can bundle only light, concave mirrors capture waves of all kinds: light, radar, radio, and even sound. (But note that parabolic mirrors are used for sound, not spherical mirrors. Parabolic mirrors possess the mathematically perfect form to focus waves, although they are more difficult to manufacture.) The concave satellite dishes on many houses receive television signals and direct them to the small antenna located at the focal point of the dish. The giant parabolic antennae, used for satellite communication, do the same on a large scale.

In the professional sphere, stars and other celestial objects are nearly always studied with concave mirrors. The necessary diameters cannot be achieved with lenses. These mirrors, which have reached dimensions exceeding ten meters (32.8 feet), are composed of hexagonal partial

---

29 This experiment only works for glasses for farsightedness; glasses for nearsightedness are fitted with diffusion lenses, which don't work like a magnifying glass.

mirrors, which can be fine-tuned to compensate for light deflection in the atmosphere.

In theory, even the death rays of the high-rise in London could be harnessed by installing a solar power plant in the British capital's financial district. Solar power plants generate energy by focusing the solar rays with countless mirrors and transforming them into electricity. The largest power plant of this kind is located in California. Mirrors with a total surface area of 2.44 million square meters (0.942 square miles) reflect light onto three towers, where steam is generated that in turn powers turbines. The output is nearly 400 megawatts—nearly a quarter of the power generated by a modern nuclear power plant. Development in this field is constantly evolving. A huge solar park that will use both photovoltaics and solar thermal power in equal measure is currently under construction in Dubai. The anticipated output? 5,000 megawatts.

Solar power plants are very much a technology of the future. Disadvantages are the huge amount of area required and, unfortunately, their impact on bird life. Desert birds should therefore avoid flying past solar towers to prevent their feathers being singed.

IMPACT SCALE

Annoying	🌧 🌧 🌧 🌧	
Lifehack	💡 💡 💡 💡 💡	
Catastrophic	💣 💣 💣 💣 💣	

Lazy Bee Against a Blue Sky

Polarized light creates beautiful effects—but caution is advised when skiing!

If you could be an animal for a day, what would you be? Perhaps a bird, to experience what it feels like to fly? Then it would be best to be a swift; they can even sleep while flying and then you wouldn't miss a minute of your twenty-four hours as an animal. Or a fish, who can breathe underwater? Our children would love to inhabit our tomcat's body to discover how it's possible to lounge around all day without getting bored. And the physicist in the house? He would love to be a bee. Not just to collect nectar and make honey, but to be able to see like a bee.

Yes, just to see. To be a lazy bee, so to speak. For bees can do something that makes physicists truly jealous: They are capable of seeing polarized light thanks to their compound eyes, which they use to determine the position of the sun for orientation, even when skies are overcast. You may wonder how this could be better than sleeping while in flight. Here's why: Polarized light is one of the most fascinating effects in physics. Without it, neither our laptops nor the digital display on our alarm clocks would work. And the blue sky in our holiday snapshots would appear much more w..shed out. To understand why this is, we need to immerse ourselves in light. Hold your breath and dive in!

We begin with a fundamental question: What is light? Generally speaking, light is the range of electromagnetic radiation that humans can see (in a previous chapter we discussed other radiation ranges that can cause windows to crack and heat the Earth's atmosphere). Illustrations in books on physics always show light waves as nice and simple, something like this:

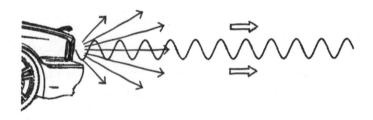

Light spreads in a linear fashion, with the electrical field oscillating perpendicular to the direction of propagation (up or down in the figure above). In physics this is called a transverse wave. You can imagine it like rope laid on the floor and then slung into the air on one end. In most cases, light doesn't just oscillate up and down but in all directions simultaneously: laterally, diagonally from bottom to top, and so forth (this is possible with a rope as well). Light doesn't have a preferred direction of oscillation. It isn't polarized.[30] Most sunlight reaches Earth in this manner, untamed and disorganized. To polarize light is to pick out individual directions of oscillation. In essence, what we want to do is to sort light—for example, by protecting our eyes from strong sunlight with sunglasses.

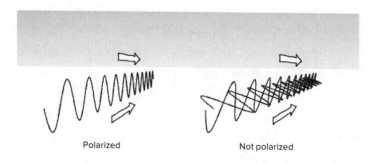

Polarized   Not polarized

Light: Tamed and trained

There are different ways of polarizing light. The simplest is to use a polarizing-filter film. Imagine a garden fence with vertical slats. A dog is running back and forth behind the fence, and you want to throw him a stick. If you throw the stick sideways or diagonally against the fence, it will ricochet. Only when you throw the stick exactly parallel to the fence slats does it land in the backyard (although

---

30 There is only one direction of oscillation that is not available to light: parallel to the direction of propagation. Light oscillates as a transverse wave always somehow perpendicular to the direction of propagation. Otherwise, it would be a longitudinal wave like sound (see the chapter "Fingernails on a Blackboard").

the dog won't need to try very hard to fetch it). This is how polarizing-filter films work. Elongated molecules in these special synthetic films only allow light of a specific polarization direction to pass through.

These films are installed in some sunglasses, for example. The lenses then only allow light to pass through that has the same polarization direction as the glasses themselves. Conversely, this means that light waves oscillating in other directions are blocked or bounce off. Therefore, the light that reaches the eyes behind the glasses is much darker than it would be without sunglasses because they aren't permeable to a large proportion of the light. We highly recommend getting these kinds of sunglasses because they allow you to constantly experience major and minor effects that the world of light tends to keep hidden from us (small polarizing-filter films, inexpensive and available online, are an alternative).

Cleaning helps

The light around us is constantly polarized even without a film. This happens whenever it is reflected from a surface or object. At home, we have parquet flooring, which performs this function beautifully (at least when it's nice and clean). Laminate flooring or tiles also work. Sunlight falls onto the floor and is reflected by it. During this process one direction of oscillation is largely lost.

Let's look at a single ray of light approaching the wooden floor. The wildly oscillating light wave encounters the electrons in the parquet, setting them oscillating in the same direction as the light wave. The electrons function like small antennae that receive the light and then reflect it. Physicists say they function like a Hertzian dipole. One way of looking at this is that the floor is full of these small antennae, which absorb the light and reflect it again.

The unique aspect of dipoles is that they don't reflect the light evenly in all directions, only in a lateral direction. Light waves that oscillate parallel to the floor are not affected by this: They hit the dipoles in the floor at a very flat angle and their light is beautifully reflected (see the top drawing on the opposite page).

The light wave oscillating perpendicular to the floor in the bottom drawing has a more difficult task: Its angle of incidence on the floor is so unfavorable that the dipole reflects the light fully back into the floor. The floor "swallows" or absorbs the light and this direction of oscillation is lost. The light reflected from the floor now only oscillates in one direction, parallel to the floor. It is *polarized*.

In nature, water and frozen surfaces play the same role as our parquet flooring. They reflect light. This can become dangerous for skiers wearing polarizing-filter sunglasses because this results in dual polarization. Imagine yourself hurtling down a mountain on your skis, heading toward an icy patch. You would usually be alerted to the ice by reflected light, which is polarized (as we demonstrated with the parquet flooring). Unfortunately, your polarizing-filter sunglasses filter out the bright reflection. This could spell trouble for your bones: As a skier, I might notice the icy and very slippery curve ahead of me too late, or not at all. Ouch! This would probably also happen to a bee. Fortunately, bees rarely go downhill skiing.

LAZY BEE AGAINST A BLUE SKY · 111

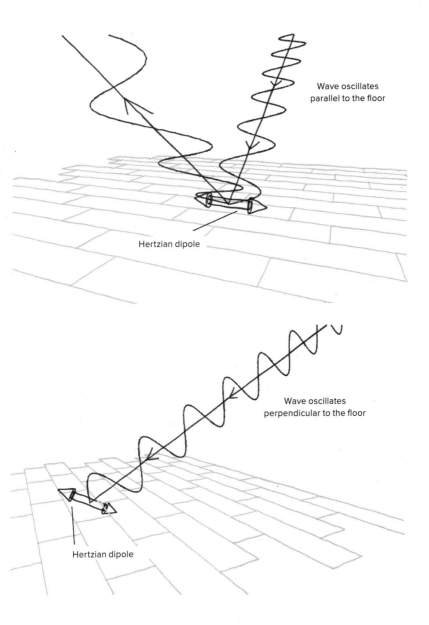

Bluer skies with polarizing filters

Should you break your leg after slipping on the icy patch, just concentrate on taking fantastic photos of the blue sky above white snow. Polarization is helpful for this as well. Most sunlight that reaches us is unpolarized, but a small portion is polarized—there are electrons in the atmosphere that function as dipoles and polarize light. Depending on the position of the sun, these dipoles are oriented in such a way that all we receive on Earth is a *single* direction of light oscillation. This happens, for example, when the sun is quite low in the afternoon. At that point, the sun, dipoles, and our line of vision form a right angle—with the result that a large amount of polarized light lands in our camera lens.

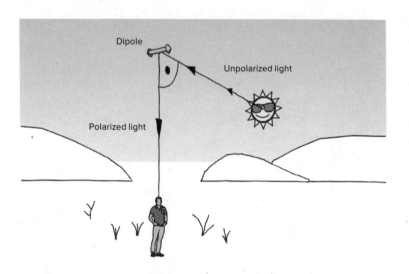

Still, this light isn't 100 percent polarized because some scattered light also penetrates from all directions. This is why photographers like to attach a filter to the camera lens (similar to the sunglasses filter) perpendicular to the direction of polarization. The filter sorts the direction of light oscillation, rendering the photograph slightly darker. The blue of the sky appears more intense, while the clouds continue to glow white because they aren't affected by polarization.

Light games in the home office

Unfortunately, we don't spend most of the year on vacation. So let us show you a fun game you can play with polarized light in your home office—a good distraction from work, should you need it. After the experiment, you will know why your laptop and your alarm clock wouldn't function without polarized light.

YOU WILL NEED:

- A switched-on screen with a TFT or LCD display (standard laptops, computers, and television screens usually have these types of display; plasma and OLED screens are exceptions)

- A switched-off cell phone (or any other dark surface that is mirror-smooth)

- Scotch tape—or any kind of transparent adhesive tape

- Cellophane as needed

HERE'S HOW IT WORKS:

Switch on your laptop or computer and ensure that the screen shows an empty, white display—for example, by opening a blank Word document. Take the Scotch tape and tape a cross or star in the middle of the screen. When we tried this, the tape didn't leave any residue when we removed it afterward. If you want to make sure, you can stretch cellophane across the screen and then stick the tape on top of that.

Now take your switched-off cell phone and hold it next to the screen. Move your cell phone around the screen once in a circular motion and look at the mirror image of the Scotch tape on the display. Can you see how the tape changes? Sometimes the screen is dark and the cross shines brightly, sometimes it's the opposite. You may even notice colors on the tape!

Take a scrunched-up piece of cellophane and hold it in front of the laptop so that it is reflected on the smooth surface of your cell phone. Now rotate the cellophane back and forth gently. You should see bright colors in the reflected image of the cellophane on your cell phone. Fascinating, isn't it?

Why does your cell phone turn bright or dark?

Screens are true polarization artists. The light they reflect is fully polarized. The same is true for many devices that display numbers or letters—digital alarm clocks, laptops, digital displays on radios and heating controls. Liquid crystal displays (LCDs) are built into all these devices. They make it possible to electronically control the rotation of the polarization direction! The displays consist of two polarizing filters on either side of a layer of liquid crystals. They rotate the polarization direction of the light when an electrical current is present.

To control the displays with precision, they are divided into different segments. Light permeates where an electrical current is present, and information is legible on the display. All other sections remain gray. In simple calculators, just seven different segments suffice to display all ten digits. High-resolution screens have millions of tiny liquid-crystal segments that render each pixel controllable, whereby each pixel also consists of three adjoining color fields.

Your screen delivers fully polarized light. This light, some of it falling through the Scotch tape, now reaches the cell phone you are tilting and moving until you hit the perfect, magical angle known as Brewster's angle. Named after Scotsman David Brewster,[31] this is the angle at which (unpolarized) incident light is reflected as perfectly polarized light. If you happen to catch the direction your laptop does *not* reflect, the reflection appears dark. If you hold your cell phone in a different position, you will see the normal mirror image because exactly the right direction of polarization is allowed to permeate the cell phone.

Pretty colorful!

There are materials that rotate the polarization direction of light. These include the liquid crystals built into your screen, as well as saccharic and lactic acids (which is why we speak of right- or left-rotating lactic acids). And, as we have seen, synthetics like Scotch tape and cellophane. This isn't visible in unpolarized light (a shame, because Scotch tape would always shimmer with a rainbow of colors). But as soon as cellophane is placed between two polarizing filters, the game is on! The cellophane acts like a switch. The light behind the first filter (the screen) is rotated by the cellophane in the perfect manner to permeate the second filter (the cell phone). When you hold the cellophane between the two filters, it renders the light visible. When you remove it, it remains dark. But the cellophane can't rotate the polarization direction of all colors equally well. This is why colors suddenly appear in the cellophane even though it is completely transparent.

---

31 Brewster was an expert in reflections and mirrors. He patented the kaleidoscope.

Even better than lazy bees

If it were possible to spend a day like an animal, another option—other than a lazy bee—would be a butterfly looking for a mate. Perhaps the Sara longwing (*Heliconius sara*), which lives in the tropical rainforests of Latin America where little direct sunlight permeates the dense foliage. The Sara longwing has a cool trick up its sleeve (or wing): It uses polarized light to attract mates. The wings of the female have patterns that reflect polarized light, which make it easier for males to locate them. The system is especially clever because birds, for whom butterflies are a popular snack, can't see polarized light. The mating signal is thus only received by willing males of the same species and the predator is none the wiser. A good thing, too: We don't want our only day as an animal to end by being devoured.

IMPACT SCALE

Electricity Hurts

**Why people get zapped
touching a doorknob but are struck
by lightning less often than cows**

The time had come for a train journey without parents or siblings. When our son was eleven, he embarked on a journey across Germany to visit his best friend, who had moved to Beeskow, a town in Brandenburg quite a distance from our home. Prior to his departure on this adventure, we talked through problems that might arise and how he should respond:

- Train delays? Keep calm and wait.

- Train stuck en route? Keep calm and wait.

- Washroom locked? Keep calm but do not wait. Walk to the next washroom.

We said a cheerful farewell at the train station and promised to turn our cell volume to "loud" so we wouldn't miss his call in the event of a crisis. We didn't anticipate any problems, especially once we received a text message shortly after departure: "All is well. There is Wi-Fi! ☺"

Two hours later, our cell phone rang. A crisis had arisen and caught our traveler off guard. We had prepared him for all kinds of glitches with Deutsche Bahn, Germany's railway network. But we'd failed to mention the headrest. "My hair's stuck," we heard him say. "It's crackling and standing straight up!" Phew! It's only something minor, we thought with relief (parents are just as nervous as their children when they send them out into the world on their own for the first time). But our son had more to say: "And the worst thing is that every time I touch a door, I get an electric shock."

Who would have thought that the most annoying problem would not be delays and packed train cars, but physics? Specifically, the electrostatic charge that causes hair to stick to headrests or static shock when we walk across a carpet and then touch a doorknob. The electrostatic discharge marks the beginning of an escalating chain of electrical phenomena that make our lives more challenging. Electric shocks, sparks, and lightning strikes can all be bothersome and even dangerous. At the same time, electricity is enormously useful, and we certainly wouldn't want to live without it.

It's not that easy to fully understand electricity. Terms like voltage, current, and output can be confusing. Let's begin with the annoying headrest and the electric charge in hair: Every body (human or object) carries electric charges, positive and negative. But we aren't aware of this because positive and negative charges usually cancel each other out. This makes us neutral. Neutral doesn't mean that there is no (electric) charge, only that positive and negative charges are balanced—much like a scale is balanced. Imagine an old-fashioned merchant's scale with two pans into which weights are placed. It doesn't matter whether the pans are empty or weighted down with an equal amount of sugar on each side. The scale is balanced provided each pan has the same weight.

Our body is only disturbed by electric charges when charged particles are added or removed. This happens all the time. Negatively charged particles, the electrons, march up our body or are discharged from it. This is because all matter consists of atoms, and these have a positively charged nucleus surrounded by negatively charged electrons. The negative charge of the electrons is balanced with the positive charge of the nucleus (just like the scale).

But the electrons are much more mobile than the nucleus. So if hair rubs against a headrest in a railway car, electrons are shifted from the hair to the headrest. Now there are too few electrons in the hair, while the headrest has too many: The headrest is negative, as it has absorbed more negatively charged electrons. The hair is positively charged because it has discharged electrons and the positively charged nucleus is overloaded.

To charge a body, an electric charge must be added or removed from it. This is a principle in physics that will become important later in this chapter when we discuss high voltage—for example, in the case of lightning. Electric charge is never, truly *never*, "created" or somehow generated in a fantastical way. It is present and simply redistributed.

Friction is a simple method of transferring an electric charge. As long ago as Ancient Greece circa 600 BCE, Thales of Miletus discovered that amber attracts small objects when it is rubbed against wool. You can try this out yourself; borrow your grandmother's amber necklace and rub it against a sweater. After that, pieces of paper or dried spices (for example) will stick to the necklace. So it's apt that the Greek word for amber, *elektron*, is the root of the word "electricity." Some also speak of frictional electricity. This isn't entirely correct because a simple touch (or contact) suffices to transfer charges from one object to another. Friction is simply a very intense form of contact!

All day long, we absorb and transfer electrons through countless contacts—when we walk, when we slide on the edge of a chair, or when we wipe a table down with a cloth. There's no avoiding these contacts, nor do they bother anyone. Just as electrons are transmitted to our bodies without us even noticing, they are also quietly transferred from our bodies when we touch objects.

This process only becomes bothersome when we are so well insulated that the charge we have absorbed cannot be transferred, as when we wear gym or running shoes with rubber soles. If we walk across carpet wearing shoes like these, we absorb negatively charged electrons. But we can't transfer this charge back to the floor because rubber isn't conductive. We aren't grounded. The charge therefore remains on us and waits for an opportunity to be transferred. As soon as we touch a material that's a good conductor, the charge seizes the opportunity and presto, the result is electric discharge. Touch a doorknob—static shock; touch a car door—a small spark will fly.

The voltages are astonishingly high: Anything above 3,500 volts is felt, but unfavorable conditions can result in considerably higher voltages. This is especially true in environments with dry air, as is the case in many offices.

The small static shocks we feel when we touch a doorknob aren't dangerous. They're just annoying. Now imagine that you're handling small electronic components, as you would be when repairing a cell phone. If a spark happens to jump onto an electrical component at the wrong moment, there's a high probability that the energy (which only slightly jolts your finger touching the component) will permanently damage the ultrathin conductors on a computer chip.

Another example is when you fill up your car. You get out, reach for the nozzle at the pump, and may well get a small shock. It won't do any harm because the nozzle, just like the doorknob, is grounded and that means you are grounded as well. Next you fill the tank. While you wait for your tank to fill up, you might sit in your car for a moment (although rules about this vary in different countries and jurisdictions), then get out again. As you do so, let's say your clothing brushes against the car seat. You then touch the nozzle again, get another shock, and a fire breaks out right before your eyes because the fuel vapor has been ignited by the spark. Luckily you react quickly and escape the flames. CCTV cameras have captured such mishaps; although rare, they can indeed happen.

Here are some tips for dealing with static electric shocks:

ELECTRICITY HURTS · 123

1. Discharge intentionally. When you touch a radiator or other conducting object several times, small amounts of charge are transferred each time. It no longer accumulates, and you won't get a shock when you touch a doorknob (or at least you'll be prepared for it). If you do this at the office, your colleagues might be slightly irritated that you're constantly checking the heating.

2. Humidify! When relative humidity is below 20 percent, a person can be charged to reach a voltage of up to 20,000 volts. That's quite a lot (even though small shocks are not dangerous). But when relative humidity rises to above 65 percent, the possible charge falls to below 1,500 volts. This is mostly caused by the very thin layer of moisture that settles on all surfaces when humidity is high, and this layer can disperse the charge.

3. Buy a carpet with metallic threads. They really exist! Some manufacturers offer such products to prevent electrostatic shock. Still, we have to admit that we don't know anyone with a carpet like this at home.

4. Use specialty tool I. It's well-known that you can buy absolutely anything online and that also goes for antistatic key chains. They look like small flashlights, the only difference being that there's no light bulb; instead they contain a metal contact. Simply tap the antistatic key chain against a doorknob before touching it. There will be a small spark (discharge)—which you can see in the tiny viewing window on the key chain—and then you can open the door safely.

5. *Use specialty tool 11.* *Don't* buy an antistatic key chain and simply use your key instead. The pointy metal tip of the key performs the same trick as the metal contact in the antistatic key chain, ensuring that the charge is safely transferred from you to the doorknob.

Unfortunately, we didn't use any of these methods when electrostatic charge derailed a major experiment for a television show. We had spent months building a three-meter-long (ten-foot) table for *Frag doch mal die Maus* (Ask the mouse), a popular German TV quiz show for children and adults. The concept was to have a very long roll of paper running nonstop across the table, like the conveyor belt in the supermarket. The goal was to determine how many meters of writing you can do with a single pencil.

Everything was ready. Marcus wanted a quick bite to eat and was about to raise the first forkful to his mouth, when his phone rang. His colleague Nils was calling; the

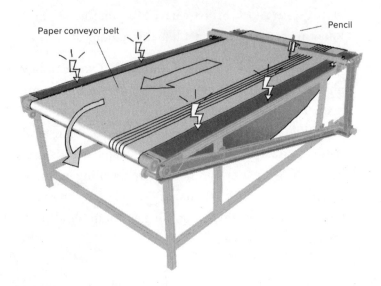

experiment wasn't working. We scrambled to find the problem and realized that the paper was stuck to the table as if it were glued there. It wouldn't move an inch. Obviously electrostatic charge was at play: There had been so much friction between the paper belt and the table surface that one side had a strong negative charge and the other a trong positive charge. There was such a strong pull between the two that the paper belt could no longer run. We had to somehow place a protective layer between paper and table.

We carefully lifted the paper off the table and brushed the table surface with our hands (after the fact, we realized we would have needed to be very careful had we then touched a doorknob, but we didn't). Then we covered the tabletop with gaffer tape: As anyone in show business knows, gaffer tape can fix just about anything. The tape has a rough surface that allows for very little friction. With just minutes to go before the start of the show, everything was working!

Why hadn't this happened when we designed and rehearsed the experiment? We don't know for sure, but we have two ideas. First, the air in the studio was much drier than the air in our warehouse. Second, there was a plexiglass wall right next to the table (a health and safety measure in the COVID-19 pandemic). The surface of this wall was covered in a thin adhesive film to prevent scratches. The film was removed shortly before filming. This, in turn, created such strong contact electrification that we could hear it crackling; it may have released so much electrostatic charge into the room that it chose to settle on our table with the paper conveyor belt. Once the

126 • THE UPS AND DOWNS OF PHYSICS

gaffer tape had been applied, the friction between paper and table was noticeably reduced and the experiment worked beautifully (and now we know that you can write for fourteen kilometers, or 8.7 miles, with a single pencil).

Electric shocks can be useful too

One of our ambitions is to find a pleasant example for each phenomenon that makes life more difficult, examples where the same phenomenon is either useful or, preferably, funny. To be honest, we doubted whether we could find an example for electrostatic charge, which seems to be bothersome wherever it occurs. But there are situations in which it is useful. Many laser printers wouldn't function without it—in other words, the precise objects that save us from having to write fourteen-kilometer-long lines with a single pencil.

Essentially, a laser printer functions as follows: Inside the printer, there is a rotating drum that will carry the image or text to be printed. This drum is electrically charged. A laser then scans the drum as it rotates, removing the electric charge in the areas exposed to the laser beam. What remains is the charge in the areas that are to be inked (in the case of toners, the ink is in powdered form). The drum then rotates past the toner, which is also electrically charged. The toner will only adhere or stick to those areas that are still charged. The drum now carries an exact image of what we want to print. As the paper slides past the drum, the toner on the drum is drawn to the paper's surface. Our document is now printed. To prevent smudging, it is fixed (or fused) by applying heat and pressure to the toner on the paper, which is why the paper is always slightly warm when it exits a laser printer. Our office printer once had a defect during this final step in the process. While it continued to print, the ink that emerged on the paper was still powdered and we could smudge or blur it simply by touching it.

Printing aside, electrostatic charges can also be very useful for cleaning. This doesn't apply to cleaning your bathroom at home, for example, but to large industrial facilities where electric filter systems (or electrostatic precipitators) are used to filter dust or soot particles from the air. Broadly speaking, electrically charged wires disperse electrons into the gases to be cleaned. There, the electrons encounter the dust (or other particulate matter) and charge it. The charged dust particles quickly migrate toward positively charged electrodes and settle there. All that is needed at this point is to cut the power and brush off or remove the collected particles.

The brutal "Current War"

Electrostatic charges may be annoying, but rarely are they truly harmful to the human body. This isn't the case for electricity from power outlets, which can be very dangerous. You've no doubt been hearing the following warnings since you were young: Never let a hairdryer fall into your bath! Don't touch an uninsulated wire! Don't stick a fork in a power outlet! These warnings are more than justified.

128 • THE UPS AND DOWNS OF PHYSICS

But why exactly? If up to 20,000 volts can be generated simply by walking across a carpet in an environment with dry air and we come to no harm, why are the 220 volts from a power outlet so dangerous?

The most important reason why a hairdryer in a bath is a bad idea is because **hairdryers work with alternating current (AC)**. You probably know that Thomas Edison invented the electric light bulb toward the end of the nineteenth century. Edison intended for light bulbs to operate with direct current (DC) so that power would flow in a circuit in a single direction, as if on a one-way street. He also wanted to earn lots of money with his direct-current patents and his electricity meters. But Edison was faced with one major problem: With direct current, a lot of energy was lost across long distances. As the electricity market continued to expand, he wanted to capitalize on this and earn additional income from the many power generating stations that became necessary. Over time, however, he lost ground to the proponents of alternating current. George Westinghouse, an inventor and entrepreneur, had joined forces with Nikola Tesla, the physics genius. Together they bet on alternating current, which changes its flow direction fifty to sixty times per second. The advantage of alternating current is that it is easily converted into very high voltages and stepped down again. It can be transmitted across hundreds of kilometers with far fewer losses than direct current. The disadvantage is that it is incomparably more dangerous for living beings. Despite this disadvantage, Westinghouse and Tesla were able to expand the sale of their patents.

Edison launched a macabre crusade against alternating current by killing animals with electric shocks in public demonstrations. The low point of this war occurred when he commissioned an assistant to build an electric chair for the government to demonstrate how deadly alternating current could be.

All to no avail: Alternating current won the day and today it powers all our electrical appliances, either through a transformer or directly, as in a hairdryer.

But what makes alternating current so dangerous? Small electrical processes are constantly playing out in our bodies. Our heartbeat is the result of one such process. Yet every heartbeat cycle contains a phase in which the heart is especially vulnerable to disruption, named the vulnerable phase or period. When an electric current or stimulus happens during this phase, the result is life-threatening ventricular fibrillation. In alternating current, the electrical impulses flow in both directions fifty times per second—and the danger that a stimulus will hit us during the vulnerable period is far greater than with direct current. This vulnerability is useful when the stimulus hits at precisely the right moment and the right strength; this is how pacemakers save lives every day.

There's another reason why we shouldn't drop a hairdryer into a bath: **Water is a less efficient conductor than we think.**

A hairdryer in water is supposedly so dangerous because water conducts electricity. We all heard this explanation from our parents when we were young. While it isn't wrong, it isn't entirely correct either. The correct explanation is this: A hairdryer in a bath is so dangerous because the human body conducts the current *better* than water. Tap water does conduct current quite well but is by no means the best conductor. Copper, for example, is one billion times more conductive. The human body is also conductive because we are composed not just of water, but also of many salts. This is why we conduct current better than our bathwater—unless we have added bath salts or had a pee in the bath (which no one ever does, of course). If the hairdryer falls into the water, the electricity spreads more easily in our body than in the water itself. This effect is further enhanced by the fact that we are reclining in the bathwater, which means that the contact area for the current is very, very large.

From bad to worse: Lightning strikes

A lightning strike is even more dangerous than the current from a power outlet. The voltage between a storm cloud and Earth is roughly 10 million volts. A current of several hundred thousand amperes can flow in lightning—and you really don't want to be struck by that. It's unclear how many people are struck by lightning in the us, for example. According to the National Weather Service, from 2009 to 2018 there were on average 27 reported deaths per year, thought to be about 10 percent of the total number of people struck (and injured to varying degrees).

Are lightning strikes more harmless than we thought? Not really, but humans have several characteristics that make them less likely victims than cows, for example.

To begin with, the flow of current in lightning is powerful, although brief. When it hits us, we benefit from the "skin effect." The current flows along the surface of the body but doesn't penetrate. (Incidentally, the "skin" in the term "skin effect" isn't related to human skin. All conductive bodies display the skin effect. High-frequency or very short, pulsed currents such as lightning run along the surface, skin-deep if you like, and barely penetrate the interior of the body.)

Secondly, lightning rarely strikes people directly and we aren't exposed to the full charge. There is a very high probability that a direct lightning strike would be fatal to a person. This can happen when standing at the highest elevation in an area, which is where lightning prefers to strike. One time, we wanted to go on a hike across tidal flats. It was drizzling and many people had brought umbrellas. Then a storm began to brew, and our guide immediately beat a retreat—there's hardly anything more foolish than standing on a tidal flat carrying an umbrella as a lightning rod in the middle of a storm. On the other hand, it's also not a good idea to seek shelter underneath a tree. Obscure sayings such as "Buchen sollst du suchen, Eichen sollst du weichen" (an old German farmer's saying, which roughly translates as "Seek out beech trees but steer clear of oaks") are complete nonsense; in a storm you should stay as far away from all trees as possible. If lightning strikes, the tree will take the biggest hit but some of it can jump to you as a result of electric discharge. This, too, can be very dangerous and even deadly. It won't help if your shoes have rubber soles; they may provide some insulation, but lightning strikes are so strong that it would simply penetrate the sole. Better to take shelter in a house, a cabin, or even a car.

If that isn't possible and you have no choice but to remain in the open, the best thing to do is to sit down and pull your legs in as closely as possible. Do not lie down! That is dangerous. Imagine lightning striking an open field (in this example, we are talking about a perfect, freshly mowed field). Here, too, the current will flow in beautiful symmetry outward in all directions. If you happen to be crouching down in the area where the current from the lightning flows out and your legs are close together, the current won't try to flow through you. Even though you may be a tolerably effective electrical conductor, the detour through your body would offer more resistance than the shorter distance through the ground.

Naturally, this will change if you spread your legs apart, or—and this would be really foolish—decide to do a push-up. Now the path through your body provides a shortcut for the current, during which it will also take a pot shot at your heart. The shorter the distance between your feet, the less the danger of something like this happening. This is also the reason why cows so often fall victim to lightning strikes: The lightning doesn't strike cows directly, but they simply can't place their legs close enough together to minimize the step voltage.

Where does lightning come from?

Where lightning comes from is a fascinating physics question. We still don't have complete explanations for the precise processes in each individual case. Broadly speaking, lightning occurs because nature doesn't like imbalances. In a storm cloud, small particles of ice collide with thicker sleet or hail pellets. The lighter ice particles strive upward, the heavier hail pellets strive downward. When they collide, electrons are transferred from the ice particles to the hail pellets. After this collision, there is a positive charge at the top and a negative charge at the bottom.

As we've just noted, nature doesn't like imbalances. It seeks to balance or offset the charge differential. This is where a fundamental physics principle comes into play: electrostatic induction, a kind of electric charge production at a distance. Electrons on the ground are repelled by the negative charge on the cloud's bottom surface. Having been repelled, the electrons move away from the relevant area, leaving behind a positive charge. Between the negatively charged underside of the cloud and the positively charged ground, a lightning channel forms—and voilà, a lightning bolt is discharged with enormous power and a temperature of roughly 54,000°F (30,000°C).

People unfortunate enough to experience this awesome natural power up close are often catapulted several meters through the air. Shoe soles are torn off, clothing is shredded, necklaces and belt buckles can melt or evaporate. As if this weren't scary enough, victims who survive a lightning strike sometimes find their skin marked with Lichtenberg figures. These branch-like patterns have been found on golf courses, a leather glove, and paving stones after lightning strikes. Luckily, the figures fade after some time.

One of the most dramatic impacts ever seen from a thunderstorm may well have been the crash of the Hindenburg airship. Together with its sister ship, the zeppelin was one of the two largest airships ever built. "Decorated" with swastika flags and carrying ninety-seven passengers and crew, the Hindenburg departed Frankfurt am Main on

May 3, 1937, en route to Lakehurst, New Jersey. The journey had taken nearly three days and was almost complete when afternoon thunderstorms broke out. The captain of the Hindenburg delayed the landing and charted a course to loop around the storm. Although the delay maneuver was successful, physics struck a blow when the Hindenburg dropped its landing ropes and mooring cable some sixty meters (200 feet) above the mooring mast, to be tied off and connected. As soon as the rope touched the ground, a hydrogen-air mixture ignited at a small leak at the top. The hydrogen was engulfed in flames and an explosion followed.

There was much speculation in the years that followed as to what had caused the fire (perhaps sabotage?). What seems logical and probable is that the Hindenburg had become electrostatically charged after flying near a thunderstorm. When the wet rope touched the ground, it was instantly grounded. The Hindenburg likely suffered the same fate as we do when we walk across carpet in shoes with rubber soles and then touch a doorknob—a trifle by comparison, and surely no cause for complaint.

IMPACT SCALE

Annoying	🌧️ 🌧️ 🌧️ 🌧️ 🌧️
Lifehack	💡 💡 💡 💡 💡
Catastrophic	💣 💣 💣 💣 💣

Fingernails on a Blackboard

Enhancing a cell phone with natural frequencies and why some sounds make us flinch

There are sounds that make your hair stand on end. Chalk screeching on a blackboard; polystyrene rubbing against polystyrene; a fork scratching across a plate—do any of these sounds gives you goosebumps? Most people find that they do.

It's a bit of an overreaction. We're not in danger when chalk screeches across a blackboard. But our brain doesn't know that. To the limbic system and the amygdala, high-pitched sounds between 2,000 and 5,000 hertz indicate danger, a primitive distress signal. One of our ancestral sisters might be screaming in fear because she's being attacked. We'd better puff ourselves up to appear more imposing to the enemy! This doesn't work, of course, because while goosebumps may make our hair stand on end, they don't make us look intimidating; they make us look like wimps. In the distant past, when we still had more body hair, it was no doubt more impressive!

What are the most unpopular sounds, the most goose-bump-inducing? Everyone has different responses. One person might hate the sound of balloons rubbing against each other, while another has no problem with it. This is likely linked to bad experiences we've had with a particular sound. And our own poll? Top of the list for us is the sound of chalk on a blackboard (and what, you may ask, does this say about our school experience?).

This sound is easy to avoid because it only happens when the chalk is held at a specific, unfavorable angle that prevents the chalk from gliding smoothly; instead, it keeps getting stuck for brief moments, then it bends a little and starts to glide again. This phenomenon is called the "stick-slip effect." Every time the chalk sticks, it bends and then relaxes again; in other words, it vibrates. We hear this vibration as a secondary noise, here a screeching or squeaking sound. We can't see the chalk getting stuck and then slipping again because it happens too quickly. However, dear teachers, please believe us: It is real! Just hold the chalk at a better angle. Or replace blackboards with smartboards and tablets in your school! Your students will thank you.

It makes you want to jump out of the car

At the opposite end of the frequency scale, there is a sound so obnoxious that Judith would rather jump out of a moving car than hear it: the dull thrumming sound when a car window is open just a tiny bit. Do you know what we mean? The wave that thrums through the car and presses down on your ears: "wwp, wwp, wwp, wwp, wwp"—it's unbearable! This usually happens when you roll down the window just a tiny bit as you start driving and then accelerate. Where does this unbearable humming come from? How can we prevent it?

Let's look at the problem on a larger scale—not in a car but in a house. You're sitting in your living room, the window is cracked open, and a large truck drives past the house. Suddenly the living room is full of alarming rattling sounds that seem to emanate from all sides. The glassware in your cabinet starts to rattle, but not because the entire house is shaking (even a large truck can't make that happen). Still, trucks have heavy engines with small explosions constantly occurring inside them, just as they do in all vehicle engines. This produces gases that expand and escape through the muffler. The frequencies of this expansion differ, however, depending on how quickly the truck is driving. When these heavy trucks start and accelerate in residential neighborhoods, they run through the full range of differing frequencies.

When the window is open (or open just a crack), the following chain of events occurs: The sound waves penetrate through the window, happily vibrate through the room, and pile up at the walls where they create positive pressure. This pressure is discharged and the air rushes back through the open window with gusto, escaping to the outside. Now there's a lack of air in the room—resulting in negative pressure. More air immediately flows into the room. This exchange happens again and again—and the glasses in your cabinet dance along in sync. A standing wave is formed in the living room.

The same thing happens in the car when we start driving with a partially opened window. At some point air flows into the car so quickly that the wavelength matches the natural frequency of the car. The partially opened window performs the same function as the labium does in a recorder. You could say that your car is playing the flute. We might argue about which is worse—a car playing flute or a plastic recorder squealing in a beginner's class—but we prefer to figure out whether we can find a positive use for this annoying effect!

Podcast from a vase

If a car can morph into a musical instrument, then surely we can use other everyday items as amplifiers! This would be very helpful because we have a problem. There's an old radio on our kitchen windowsill that for many years has entertained us while we wash the dishes. Its days are numbered, however, because two things have changed. First, we now have a cat who prefers to leave and enter the house via the kitchen window. To this end, we need to move the radio from the windowsill. If we don't, our cat brushes against it and shoves the radio into the sink (on purpose!). Consequently, someone in the family always has to move the radio from the sill, accidentally pulling the plug and erasing all station settings, which annoys the next person to do the dishes because they don't notice until their hands are wet and covered in suds and they can't easily reconfigure the stations.

The second change is that we are in the grip of podcast mania, and podcasts are broadcast via our cell phones, not the radio. Listening to podcasts is a little bit like listening in on a conversation at the table next to you in a café—at times fascinating, at times funny, and nearly always very long. The longest episode of our favorite podcast ran for seven hours and thirty-nine minutes. At such a length it doesn't matter if you can't hear every word because you're washing dishes, vacuuming, or driving.

But then we discovered a finance podcast. The topic was thermostats and how to get rich from them—totally niche and all the more fascinating for it. As is de rigueur for a financial topic, the guest spoke in a calm, sonorous voice without much modulation. This became an issue, especially when we were driving. We couldn't hear a word of the podcast because our car radio had been installed prior to the invention of Bluetooth. We quickly ran through the options: buy a new car (too expensive), pull over and finish listening to the podcast (too long), or build a better loudspeaker.

Jell-O and a pencil

From the perspective of physics, a loudspeaker fulfills one principal function: to convert electrical signals into sound—that is, into oscillations in the air. This is difficult when the loudspeaker is as small as that of a cell phone. Imagine the air as a giant Jell-O and the membrane in the cell phone loudspeaker as a sharp pencil. Now you want to make the

Jell-O jiggle by pricking it with the pencil. This will barely elicit a response from the Jell-O; at most it might move a little. It would be more effective if you used the thick eraser at the end of the pencil. Or something even larger, a potato masher, for example.

Large, high-performance loudspeakers therefore also have large membranes. The larger the membrane, the easier it is to transfer a lot of energy to a great deal of air—that is, to make the air vibrate more strongly and give the loudspeaker more oomph.

At that time, there was no large membrane in our car to which we could connect the cell phone. That left us with only one option: We had to direct the small amount of sound emitted by the cell phone to where we wanted it to go. Specifically, the sonorous voice of the thermostat millionaire needed to reach our ears. If we just placed the cell phone on the passenger seat, the sound from the small speaker would spread in all directions and only a small portion would reach us.

We wanted to change this. One advantage for us was that sound waves don't move easily from one medium to another. A concrete wall, for example, provides fairly good protection against hearing conversations in a neighboring apartment; most of the sound is reflected at the wall and can't easily penetrate concrete. We deliberately choose to say "not easily" because concrete walls aren't fully soundproof. Just ask our neighbors whose living room is on the other side of the wall where our piano stands—they could tell you a thing or two about it. But if the wall were solid wood, the piano sound would be even louder next door. And there's a good reason why no one plays piano in a tent.

Why is it so loud in a tent?

To answer this question, we need to understand what sound is—namely, vibrations. Vibrations of this kind can occur in any medium—in water, in pudding, and, naturally, in air. In the cell phone's speaker, an electrical impulse nudges a few air molecules. These are set into motion and nudge the adjacent molecules and so on and so forth. By this means, sound travels through the air in a kind of chain reaction. Picture a metal or plastic Slinky, the popular children's toy that can travel down a set of stairs; when you pull the Slinky slightly apart and then nudge it at one end, the energy of this nudge moves like a wave through the coils.[32]

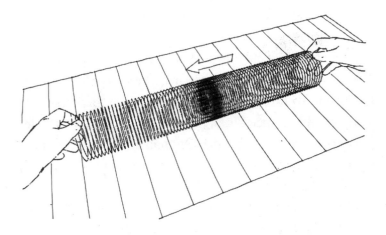

---

[32] For fans of details: If you nudge the Slinky in the direction of its coils, we speak of a longitudinal wave. If you nudge it sideways, the coils bulge sideways. This is called a transverse wave. In air, sound only travels as a longitudinal wave. See this excellent short video created by the National Music Centre (Studio Bell) in Calgary, Canada: "Slinkys and Soundwaves," posted September 6, 2015, YouTube, 2:18, https://www.youtube.com/watch?v=kxQj-wPePBU.

142 · THE UPS AND DOWNS OF PHYSICS

Sound also has another property that's very useful to us in constructing an improvised loudspeaker in our car. To put it bluntly, sound is a bit lazy. It doesn't like to switch from one medium to another. If you've ever dunked beneath the water in a bathtub while someone is standing next to the tub and speaking to you, you'll know what we mean. Sound from the air doesn't like to dive into water. Underwater, all you hear is the muted sound of the voice.

Physicists like to measure things. To calculate how well sound travels from one medium to another, they introduced a term: (acoustic) impedance. *(Caution: We're about to delve into truly complex spheres of physics. Persevere, you'll be proud you did!)* Two factors play a role in impedance: the density of a material and the speed with which sound waves travel through the material.

Air has relatively low density; after all, it is a gas. Sound doesn't travel quickly through air—it moves at a speed of roughly 340 meters (1,115 feet) per second. It travels much more quickly through other materials! In synthetics, sound can travel at a speed of 2,300 m/s (7,546 ft/s). Synthetic materials have a higher density than air. The impedance of air and that of synthetics are therefore very different. The greater the difference in the impedance, the less likely sound is to transfer from one medium to the other. In other words, sound doesn't like to travel from air to the synthetic material.

This is very practical for constructing our loudspeaker! Imagine that our podcast sound wave is traveling at a leisurely pace through the car. Suddenly it collides with the dashboard, which is made of synthetic materials. What happens to the sound wave now? The same thing that happens to us when we're forced to hit the brakes and aren't wearing our seat belts: It collides with the dashboard and bounces back from it. At the very least, this would be painful for the human body, but it has hardly any impact on sound. If a synthetic material can reflect the sound, it should also be capable of directing it to our ears. And this is indeed what happens: The tray beneath the windshield in our car has a small compartment for sunglasses. If you place the cell phone inside the compartment, leaving the cover open, the sound is reflected by the cover and lands in the driver's ears. (If you don't have a compartment for sunglasses, you can place your cell phone in the compartment in the center console. The sonorous voice may emanate from a lower point, but it's still very audible.)

Cardboard doesn't make a good loudspeaker!

This DIY loudspeaker always works well when the impedance of the material is distinctly different from that of air. Glass and porcelain work well; cardboard and Styrofoam do not. By the time we reached home, we knew all there is to know about thermostats thanks to our DIY loudspeaker. But we really don't care that much about thermostats. What we really want to know is what is the best household vessel to create the best cell phone amplifier at home.

Test object 1: Our largest vase. Almost knee-high, it usually holds sunflowers. The podcast starts up; while the voice from the vase does indeed sound louder, it is also quite muffled. Test object 2: A somewhat smaller vase, with a slender middle and a wide opening at the top. It sounds quite good. And on we go, trying out one vase after another. It's soon evident that size doesn't matter. Taller and wider vases don't necessarily amplify the sound; they often sound worse than the smaller vases.

Two hours later, we have a winner: Our smallest vase. It stands some fifteen centimeters (six inches) tall and has a diameter of twelve centimeters (4.7 inches). While the volume is as loud as with the larger vases, the tone is much clearer. It's even louder when we hold the loudspeaker of the cell phone to the bottom of a vuvuzela that has been lying around in our basement since the World Cup in South Africa. This method was quickly eliminated, however, because it takes two hands to hold the vuvuzela next to the cell phone and considerable contortion to position your ears anywhere near the speaker. Besides, it wouldn't leave our hands free to do the dishes.

The sound of empty mugs

What does the small vase have that the large ones don't? First, it has a helpful inverted-conical shape. The base has a small diameter, and the vase widens toward the top. Thanks to this funnel shape, the surface from which the sound is reflected also widens, constantly amplifying the volume (bugles, trumpets, and trombones look the way they do for a reason). The funnel shape serves to transfer the energy that has entered at the bottom or narrow end to the air in the most effective manner.

Second, the small vase is the perfect size. Every object has its own natural frequencies in which the air contained within it vibrates, depending on shape and size. When I place my cell phone in the thick floor vase, different pitches are amplified than in a juice glass.

By the way, you can hear these differing natural (or resonant) frequencies if you place your ear on top of an empty mug or glass. Perhaps you have a mug of coffee or tea next to you at this very moment. As soon as the mug is no longer full of a hot beverage, place your ear on the opening: At first you will hear a rushing sound similar to that from a conch shell at the beach (when we were children, we thought this was the sound of the sea until we were told, also incorrectly, that it was the blood rushing in our ears). Keep on listening—then you'll notice a sound that may differ depending on the container. You can even sing along to it.

Why are you hearing a sound? The mug itself doesn't make sounds—and (hopefully) you don't have tinnitus. What happens is that the ambient noises around you flow into the mug, short and longer waves, a jumble of all kinds of frequencies. But only the noises with a specific wavelength stimulate vibration in the mug because they hit exactly the right frequency, which is the natural or resonant frequency of the mug itself.

It's the same as with the standing wave in the living room. A pressure wave spreads in the mug, and the air briefly accumulates at the bottom. This creates momentary positive pressure. This pressure is released and pushes the air upward—and it does so at great speed, the speed of sound. Because our mug has a large opening (we want something comfortable to drink from), the air can travel freely; it is inert, like all masses, and once it is set in motion it doesn't easily stop. More air escapes from the top than was in the mug in the first place. Instead of positive pressure, there is now negative pressure at the bottom of the mug. The ambient air offsets this negative pressure in that new air quickly flows into the mug—once again accumulating at the bottom and creating positive pressure. The pressure wave keeps oscillating back and forth, and it does so in the precise rhythm compatible with the mug.

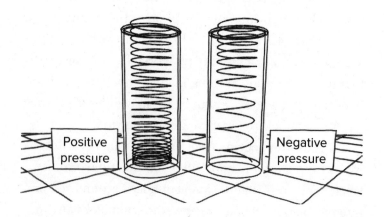

The height of the mug determines how the air vibrates back and forth. If the mug is too tall, the pressure takes longer to release and recharge. If the mug is short, the same process occurs more quickly. The air vibrates in the natural frequency of the mug—or the vase. Sounds that lie within this range are properly amplified and sound good.

Podcast yes, music no

Our winning vase, fifteen centimeters (six inches) tall, is suitable for amplifying spoken words. Even a smaller container such as a wide-rimmed glass would work well (all kinds of creative DIY cell phone amplifiers are available online made from tubes, mugs, or glasses). They amplify the frequencies of the spoken word.

It only works for language, mind you! Our daughter took the winning cone-shaped vase into the bathroom to listen to her favorite band loud and clear while she was having a shower. The music was indeed loud, but it sounded awful. Dull and tinny—in short, unpleasant. While our thirteen-year-old was terribly disappointed, the physics behind this result was fascinating! The vase loudspeaker isn't suitable for music because it contains too many different frequencies. If you want to listen to music with good sound quality, you need to get the large floor vase from the basement. Or turn on the old kitchen radio when you're doing the dishes. The radio will keep occupying its place of honor in the kitchen for listening to music. It has such a lovely loudspeaker that makes everything sound wonderful. And the cat? Well, he can step around it!

148 · THE UPS AND DOWNS OF PHYSICS

IMPACT SCALE

Annoying

Lifehack

Catastrophic

Foggy Glasses and Cloudy Mirrors

~~~~~

**How to outsmart humidity
and see all the better for it**

Humidity—what a boring topic, you may be thinking! But solve this riddle first: Which do you think is heavier, humid air or dry air? The solution is at the end of the chapter.

Let's start by looking at something truly bothersome. Foggy glasses are particularly annoying when you come inside on a cold winter's day. A common occurrence we're all familiar with and, until recently, hardly worth mentioning in a book. Then came the COVID-19 pandemic. And with the mandate to wear face masks, foggy glasses morphed from niche topic to universal problem. For when you wear a mask, glasses fog up all the time, not only when it's cold outside, but simply because our humid breath is directed upward by the mask.

Our son, who was spending eight hours a day in school wearing a mask and glasses, was especially annoyed. He took to pushing his glasses down to the tip of his nose, but that didn't help either. He chose a mask with a metal nose wire, fitting the mask tightly to his face—alas, without success. He purchased anti-fog spray but didn't derive any result other than smudgy glasses. In the end, he simply took his glasses off in class (which wasn't good for his eyes or for his grades, because now he could see the blackboard but couldn't always read the writing on it).

Our son's constant griping made us aware of just how annoying humidity can be, this (water) vapor in the air that blends with the other components of the air we breathe, such as oxygen and nitrogen. How can we come to grips with it? Can physics help us outsmart foggy glasses?

The root of the nuisance is that warm air can hold more water vapor than cold air. Beneath the mask, close to our face, the air is very warm (= a lot of humidity). On top of that, we breathe out inside the mask (= even more humidity). This warm, humid air rises through the small gaps between mask and nose and hits the glasses, where it is colder because the glasses are in contact with the cooler air that surrounds us. The air we breathe out cools at the glasses, where it reaches its dew point. This is the temperature at which air can no longer hold water, to put it simply. Moisture begins to settle on an object, in this case on the lenses of our glasses.

We experience the same phenomenon on a slightly larger scale with bathroom mirrors. After a shower, you can't see anything in the mirror. Seems illogical at first. We've heated the bathroom nice and cozily and warm air can hold more water vapor than cold air. It's true that it can hold more—but it can't hold infinitely more. When we shower, an extreme amount of water vapor is discharged into the air, much more than even heated air can handle. At some stage the dew point is reached, and the relative humidity lies at 100 percent. This means that the air can no longer absorb any vapor. The vapor condenses on the mirror, the tiles, or the window. Similarly high humidity levels are otherwise only found in tropical rainforests, where humidity is consistently between 90 and 100 percent.

What is bad air?

A steamed-up bathroom or humid tropical rainforest isn't a pleasant space to spend time in. Humans tend to be comfortable in air humidity of between 50 and 60 percent; this is sufficient to keep the mucous membranes in the mouth and nose moist. At the same time, air at this level is dry enough for us to expel water, something we are constantly doing even though we may not always be aware of it. When we engage in sport we sweat, of course, and we notice it. But we transpire all the time, even when we're not exerting ourselves. We expel roughly half a liter (0.1 gallons) of water into the air through our pores and an additional half liter through our breath (this explains why it's a good idea to drink plenty of fluids).

Take our situation: The bodies of the six members of our family expel a total of six liters (1.6 gallons) of water into the air in our apartment every day. Can the air even absorb all that humidity? Physics has two methods to find

152 · THE UPS AND DOWNS OF PHYSICS

out. We could lock ourselves in for three days, opening neither windows nor the front door during this time. Or we can calculate it.

Even in the days of COVID-19 lockdowns, the experiment seemed too onerous, so we decided to go with the calculation. Our living room has a floor area of 20 square meters (215 square feet) and a ceiling height of 2.4 meters (7.9 feet), which results in forty-eight cubic meters (1,695 cubic feet) of air. Now we need to know how much water vapor a single cubic meter of air can absorb. As we have seen, this depends on the temperature. The warmer the air, the more water vapor is needed to saturate it. Imagine you're standing in a desert with a temperature of approximately 95°F (35°C), with 7.6 milliliters (0.26 fluid ounces) of air in a cubic meter. You would likely experience the desert air as very dry—justifiably so, because this value corresponds to a relative humidity of only 20 percent. In other words, the air has only absorbed 20 percent of the water volume it is capable of absorbing. But that is all the moisture that is available in the desert.

As night falls, the temperature in the desert drops radically. Large temperature swings are not unusual in the Sahara, for example. Let's assume that the temperature in the desert has dropped to roughly 45°F (7°C). There is still 7.6 ml of water per cubic meter of air. But the air has cooled to such a degree that it can only absorb less water; it barely manages the 7.6 ml. This is why the relative air humidity is suddenly at 100 percent even though the amount of water is the same! Incidentally, this is how water can be collected in the desert. In the evening, a thin film is spread out on the ground and condensation from the cooling air is collected in the early morning hours. If you cool the inside of your car with the air conditioning in the summer, you'll end up with lots of condensation on the windshield. Luckily, air conditioning manufacturers have taken this into account: The condensation is funneled away in a hose and the car is comfortably cool with equally comfortable air humidity.

Whether or not water condenses therefore depends not only on the absolute amount of water vapor in the air, but also on temperature. When mist hovers above a meadow in the morning, the temperature at the meadow is still cold enough that the air needs to "park" some of its humidity there. As soon as the air warms, it absorbs the moisture, and the mist disappears.

Freezer compartment— an arctic desert

This effect is clearly visible in the freezer compartment of our fridge. It constantly looks like an arctic desert; the interior walls are covered in ice crystals, forgotten packs of fish sticks frozen like fossils. The fridge, including its freezer compartment, stands in the kitchen, the warmest room in our house. This is where we cook and bake, and all these activities create so much warmth that we rarely need to heat the kitchen. But both activities also create a lot of humidity. Cooking produces up to 1,500 ml (51 oz) of water vapor per hour, which is even more than a shower (up to 800 ml, or 27 oz). When we open the freezer compartment, warm, humid air from the room flows into the icy compartment—and immediately condenses into irksome

154 · THE UPS AND DOWNS OF PHYSICS

(albeit beautiful) ice crystals. The only solution to this problem is to no longer open the compartment. That would presumably mean that there would be no ice crystals—or perhaps not? We wouldn't be able to tell because we wouldn't be opening the compartment. The situation would be akin to Schrödinger's famous cat...

But let's get back to our living room. At a temperature of 68°F (20°C) and a comfortable humidity of 60 percent, each cubic meter of air contains 10 ml (0.34 oz) of water. We have forty-eight cubic meters (1,695 cubic feet) of air in the room, so there are 480 ml (16.2 oz) of water coursing through the space. That's nearly half a liter. If we are joined by friends and there are four of us sitting in the living room playing a game, each of us expels approximately 50 ml (1.7 oz) of water vapor per hour (for the purpose of this calculation, a "game" is defined as moderate physical activity; we're not playing dodgeball). This translates to an additional 200 ml (6.8 oz) of water vapor per hour and a constantly rising relative humidity, which isn't a problem for the first two hours because we started at 60 percent humidity. But if we remain seated for a long time and don't air out the room, it will become uncomfortable.

We don't need technical instruments to measure humidity; we can easily feel it. When we are all together, someone will frequently ask for the room to be aired for a while. If we didn't do that, the relative humidity would eventually reach 100 percent. The humidity would settle in the room— on furniture, on clothing, on the walls and windows. And if we continued that way, at some point we'd have mold. Mold loves humidity and thrives in settings where relative humidity remains above 80 percent for a longer period.

In other words, the humidity we humans create must be removed. This is the chief purpose of regular ventilation: letting out carbon dioxide and removing humidity. Of course, this only works if it isn't even more humid on the outside. In summer, for example, warm external air brings a lot of humidity into a room because it contains more water vapor. In summer it is therefore best to open windows in the evenings or early in the morning when the air is still cool; it will have deposited part of its water vapor as dew on the lawn and won't carry it into our home.

The right way to ventilate

In winter it doesn't matter; you can open the windows to let air in throughout the day. Still, in winter we tend to keep the windows shut—after all, it's so cold! Our family is divided into fresh air fanatics and T-shirt wearers. The former want to open all the hatches wide several times per day, regardless of how cold it is outside. The latter are mostly interested in not feeling cold. A typical morning in our household plays out as follows: Judith is the first in the kitchen and opens the window wide. Marcus arrives a minute later and closes the window. Our daughter shows up and opens the window again, whereupon her brother complains—and on it goes. The T-shirt team can argue that every airing naturally decreases the interior temperature. This means that it must be heated again, which costs money. This is why experts recommend letting in lots of air in short bursts and then closing the windows again. In other words, don't leave windows open a small amount over long periods of time. Unintentionally, our chaotic morning

The father of bad air

Whether a room feels comfortable isn't just a question of humidity. There are other components in the air, such as oxygen and carbon dioxide, and the concentrations of these components determine whether the air in a room is "bad" or not. The scientist Max von Pettenkofer studied these factors intensively as early as 1858.[33] You could call Pettenkofer the "father of bad air." He focused on hygiene and its effect on human health. He was also a fan of experiments. Famous among them is his self-administered test of swallowing a sample laced with cholera bacteria to prove that the bacteria alone couldn't cause the disease. He was lucky and suffered only mild symptoms and some diarrhea.

Pettenkofer sealed a room as much as possible and observed the air composition in which his test subjects felt comfortable. This is how he discovered that people prefer less than 0.1 percent of carbon dioxide in the air, which corresponds to 1,000 carbon dioxide molecules per 1 million air particles. At this volume, Pettenkofer discovered, the odor generated by other people in the room was at its lowest nuisance factor. This "Pettenkofer factor" was the standard measure for good air quality for a long time and has only recently been replaced by more differentiated measures.

---

33 Max von Pettenkofer, *Über den Luftwechsel in Wohngebäuden* (Munich: Cottaesche Buchhandlung, 1858).

Pettenkofer didn't use ventilation in his experiment. He assumed that the walls contributed considerably to air exchange. In experiments on "breathing walls," he sealed a single face of bricks and discovered that air could still permeate. He deduced that air exchange through walls was a significant factor in good interior air quality. This conclusion isn't quite true[34] because Pettenkofer was working with very high pressure, which isn't present in everyday life in our contemporary apartments. What is notable, however, is that even all those years ago, he noted that air exchange was less efficient in the walls of modern houses. The same is true today: The better a home is insulated, the more we must ensure that humidity doesn't result in interior condensation. A well-insulated house keeps the outside on the outside and the inside on the inside.

If you're annoyed by constant ventilation, heating, or cleaning your glasses, you may wish to consider the positive aspects of condensation. For example, without it there would be no clouds. Clouds form because air cools at great heights. It can then no longer hold the humidity, and the vapor condenses. Those things floating above us in the sky are massive amounts of water! Depending on the size of the cloud, that's roughly one hundred tons of water, equivalent to the weight of twenty elephants. At some point, this weight falls on us: When the cloud becomes too heavy, rain falls. This is clearly a positive aspect of humidity because we need rain (given the extreme heat in recent summers, we've become even more aware of its benefits).

---

34 H. Künzel, "Kritische Betrachtungen zur Frage des Feuchtigkeitshaushaltes von Außenwänden," *Gesundheits-Ingenieur* (1970).

Besides, humid air is also very useful for mounting a spectacle! In our TV science show, we perform an experiment in which we spray liquid nitrogen several meters into the air. The nitrogen evaporates, the air cools considerably, and the vapor contained in it condenses into a cloud. The responses to this effect range from "Pretty cool" when the air is on the dry side, to "Wow, amazing!" when the air is humid and a dense cloud forms which then sinks slowly to the floor. A similar effect happens when you take ice cream out of the freezer and a "tail" of fog will often flow down from the ice cream. The more humid the air, the more pronounced the effect. When the air is completely dry, no fog is visible.

Heatable glasses and other tricks

Although we can't prevent irritating condensation in inconvenient places, we can at least outwit it with a few tricks. One possibility is to warm the air at critical spots so that it can absorb more humidity. For example, I can blow-dry the mirror in my bathroom. It takes a bit of time, but it works. Alternatively, I can wipe the mirror with a towel, although it usually just fogs up again. We encountered a particularly clever solution on a business trip to Japan, where we found a small heating coil mounted behind the mirror. This coil heated a letter-paper-sized surface area, which meant it was no longer one of the coldest surfaces in the room and hence remained free of fog or condensation. If you want to obtain the same result with a towel, you just have to rub the mirror vigorously to warm the mirror glass sufficiently and stop it from fogging over again.

FOGGY GLASSES AND CLOUDY MIRRORS · 159

Unfortunately, this great idea doesn't work with fogged-up glasses (although heated frames might be an option... ?). Instead, you can try the soap trick: Take a dry piece of soap or a drop of dish soap and carefully rub it onto the lenses. Then polish your glasses with a soft cloth, ideally a cloth designed for cleaning glasses or a microfiber cloth, taking care not to scratch the glasses. This should improve the situation because the soap leaves behind a wafer-thin film. Although this film doesn't prevent condensation, no droplets will form, only an even film of water. The soap decreases the surface tension of the water, thereby preventing water droplets from settling on the glasses. The same approach also works for bathroom mirrors. In both cases, the soap film needs to be reapplied every few days.

And if you don't have a bar of soap or a drop of dish soap to hand, you can make use of a magic biological ingredient: spit. You might be familiar with this trick from the pool, where swimming or diving goggles fog up for the same reason as described above: The air trapped inside the goggles gets gradually more humid and condenses on the cooler surface of the goggles. If you spit on the goggles and then rub the spit around, the problem is solved.

Although our spit consists largely of water, that isn't its only component. It also contains proteins, which are very helpful. These proteins are called mucins. In principle, mucins are simply mucus. They are produced by plants as well as animals and humans. For example, mucus helps our food to slide down the esophagus to the stomach. Not terribly appetizing, but very practical!

So when you spit on your diving goggles, the proteins spread across the lens and can't be easily washed away with water (try to clean a plate covered in the remnants of a protein-rich creamy cheese sauce without detergent). As a result, the water droplets formed by condensation pearl and simply run off, leaving you with a clear view. We gave our son the same tip to deal with his glasses/face mask problem. We weren't surprised when he rejected the idea ("I'm not going to spit on my glasses!"). But from the perspective of physics, we know that it would have worked.

Now that we've reached the end of the chapter, we still owe you the solution to the riddle we posed at the beginning. Which is heavier: dry or humid air? You may have thought it must be dry air based on how we've formulated the question. And you are correct! In proper physics terms, dry air has more density than humid air. At first, we were surprised by this fact, but after a little bit of thought it is all too plausible.

Let's take air saturated with humidity at a temperature of 68°F (20°C), where one cubic meter of air would contain 17.3 grams (0.038 pounds) of water molecules,[35] which whiz around between the other molecules in the air. While air humidity is but a small part of air, every forty-third particle in air is a water molecule. The other molecules are nitrogen, oxygen, and argon. We can disregard the trace gases (carbon dioxide etc.) here.

Now we will dry our humid air—for example, by setting up a dehumidifier. Practically speaking, this means that all water molecules—that is, every forty-third particle—

---

35 Given an atmospheric (or air) pressure of 1,013 millibars (1,013 hectopascals or 1 atmosphere), which corresponds to standard ambient pressure.

disappears from our air volume. To equalize the pressure, air particles flow in from the outside. In the end, every water particle is replaced by an air molecule. We can actually picture the process in this fashion because the volume of the gas is nearly independent of the type of molecules or particles it contains as long as the number of molecules remain the same.

A water molecule weighs 18 u ("u," or Da for dalton, is a unit of measurement for the mass of atoms and molecules). An air molecule is heavier; on average, it weighs 28.9 u. So once we exchange our lighter water molecules for their heavier colleagues, dry air must indeed be heavier than humid air. For a cubic meter of air, this difference is roughly ten grams (0.02 pounds).

Despite all this, the warm humid air on a summer's day feels heavy. The humidity settles on our skin, and it becomes more difficult to transpire. This is very uncomfortable, and knowing that dry air would be even heavier offers little comfort.

IMPACT SCALE

Annoying

Lifehack

Catastrophic

RIP, Cell Phone!

Why diffusion gives iPhones a horrible death but delivers crunchy carrots

When you breathe in helium, your voice sounds like Mickey Mouse: high-pitched, squeaky, artificial, and hilarious. Helium is a light gas in which sound spreads more quickly than it does in air (only hydrogen has even lower density). This is what makes your voice sound so high-pitched. It's so much fun, and we sometimes let our children breathe in helium when they visit us in our workshop, although only a little bit and in a very controlled manner. As a light gas, helium dissipates from the lungs on its own. But if you breathe in too much of it, you may end up with too little air left in your lungs. Some people have fainted, fallen, and bumped their heads. So be careful when breathing in helium!

For a television program, we were asked to find a safe way for celebrity guests to inhale helium. The great German comedian and presenter Wigald Boning was among them. The plan was to build a walk-in box containing a healthy ratio of helium and oxygen. The idea was that Mr. Boning and his colleagues would step into this box. The fact that helium is such a light gas made things easier; we could simply leave out the bottom of the box and elevate it so that Mr. Boning could slip in from below. After all, the helium floated nicely at the top.

The box was a complete success! Wigald Boning sounded like Mickey Mouse and we were satisfied, as was the broadcaster—until Mr. Boning spoke to us (after the show, once his voice had returned to normal). His cell phone was broken after being in the box. It was an iPhone 6, a brand-new model at the time, so this was very annoying.

We were stumped—after all, we weren't aware of doing anything wrong. Helium is a noble gas and extremely inert. It doesn't react with oxygen, doesn't burn, and doesn't form compounds with other substances. What's more, during

164 · THE UPS AND DOWNS OF PHYSICS

rehearsals we had also been in the box with our cell phones without any damage to people or equipment. One of our employees even had an Apple device that had survived the helium bath undamaged.

All we could do was respond in a friendly manner and say that we couldn't understand why the damage had occurred. We put the matter behind us but kept it in the back of our minds as a curious anecdote. A few months later, we happened to come across a report about Erik Wooldridge, a systems specialist at Morris Hospital near Chicago. In 2018, he had just installed a new MRI scanner at the hospital when the doctors and nurses approached him: Cell phones that went near the scanner were breaking down in droves. And it wasn't just cell phones; smartwatches were also affected. Erik Wooldridge's immediate and terrified thought was that the MRI was emitting electromagnetic radiation. That would have been a huge problem! But if that were the case, cell phones wouldn't be the only things affected; other medical devices all around the MRI would also have been impacted. There were quite a few of them, and they were all doing fine.

Wooldridge looked at the damaged cell phones and realized that they were all Apple devices—iPhones and Watches, forty in total. What could have caused this? Wooldridge posted the problem on Reddit, where other system administrators quickly suggested that it could be due to the helium used to cool the MRI. Several hundred liters of liquid helium are needed in these devices to cool the superconducting magnets. And indeed, Wooldridge discovered a small helium leak! However, this still didn't explain why the iPhones reacted so poorly. When he lined up the malfunctioning phones, a clear pattern emerged: The newer the model, the worse the damage. All the affected iPhones were model 6 and higher (all models of Apple Watch were affected). The only iPhone 5 still in use on the ward worked perfectly.

This caught our attention, so we asked our employee which device he was using; it turned out to be an iPhone 5, apparently just old enough to survive the helium bath, unlike Wigald Boning's new model. Of course, now we were faced with another question: What's different about the newer devices? What makes them so vulnerable to helium?

Why are only newer iPhones affected?

If the Men's Eight rowing team is going to race, strong rowers are essential. But so is a coxswain to set the pace. Every computer and smartphone also has a timer that sets the pace; in this case, it's called an oscillator. The oscillator receives small electrical impulses that cause it to vibrate. The frequency of this oscillation determines the frequency of the computing steps in the cell phone's processor.

Now imagine if the coxswain of the Men's Eight were to get drunk before a race. He would no longer be able to keep time; first he would count far too fast, then slump backward and fall fast asleep. The entire Men's Eight would be thrown into disarray, especially when the coxswain suddenly started counting twice as fast, and the rowers would soon become exhausted. This is what happens to iPhones when they are exposed to helium.

Why is it so easy to damage these iPhones? The short answer is that they're a bit delicate. This is due to the type of oscillators that Apple uses. In most modern computers, small quartz crystals perform this task. These are small blocks that expand and shrink very quickly under voltage. They are stimulated to oscillate by electrical impulses. This is a great piece of technology that can keep time precisely.

Unfortunately, quartz crystals have a few disadvantages. They are relatively thick, and they are also very sensitive to heat, cold, dirt, moisture, and vibration. The crystals must be protected by ceramic housings, for example, which in turn are expensive and time-consuming to produce. What's more, we all want our smartphones to be practical and slim. Like all other manufacturers, Apple looks for components that are as small as possible.

Apple found its solution in MEMS chips (or the more unwieldy "micro-electromechanical systems"). These are tiny components (the entire side length is just one millimeter, or 0.04 inches) in which even tinier silicon lamellae (or wafers) oscillate back and forth. These wafers are so small that they can only be seen properly with an electron microscope. MEMS oscillators have many advantages over quartz. They are more precise, cheaper, more resistant, and less sensitive to cold. However, as you may have guessed, their Achilles heel is helium. While a quartz crystal is not affected by the presence of any gases, in MEMS oscillators the helium atoms simply migrate into the chip and do so astonishingly quickly. In tests, the phones stopped making a sound after just four to eight minutes.

Water no, helium yes

Most smartphones may be waterproof, but gases can still get in, especially if it's a light gas like helium. The physical process behind this is called diffusion. Simply put, diffusion means that the particles in gases or liquids mix until there are equal numbers everywhere. This is very practical, as it also distributes oxygen throughout the air we breathe. Imagine if you ended up standing in a spot with pure oxygen. That would be bad.

The Scottish botanist Robert Brown discovered the mixed particles in 1827 when he was observing pollen and dust under his microscope. He saw that particles were constantly moving and dispersing, a type of movement later called "Brownian motion." Albert Einstein built on this when he published his seminal work on molecular kinetic theory in 1905. Einstein concluded that there must be tiny, invisible particles in liquids and gases that push the pollen back and forth. Brownian motion is therefore proof that atoms and molecules exist and that they are constantly moving.

Back to Wigald Boning: As he climbed into the helium box with his new iPhone, there was a lot of helium around the cell phone and none in the oscillator chip inside it. This is because there is normally a vacuum there (these components are produced in an atmosphere of hydrogen. They are then baked at low heat, the hydrogen escapes, and the vacuum remains). Helium quickly diffused into the chip to compensate for the concentration gradient.

The sudden influx of helium "confused" the oscillator. The components no longer oscillated in the vacuum but in a thin gas. The frequency changed. The electronic component that controls the oscillation reacted randomly. The cell phone's timer now called for slower—or faster—frequencies and at some point it simply ceased to work. Tests showed that the timer (or oscillator) of a cell phone runs faster at first, then slower, and finally stops altogether.

Practically speaking, the risk of this occurring is low for most iPhone users, few of whom will be working near a source of helium. Apple's decision to create an oscillator made of silicon is therefore understandable. Moreover, Apple handles the issue proactively and even mentions the problem in its user guide. They recommend that if a cell phone has fallen into a "helium coma," you should simply let it rest for a few days. After a week, the helium will have dissipated from the oscillator chip.

Fun with diffusion

If you've been affected by this and are waiting for the gas to dissipate from your cell phone, you could use the time to carry out a few fun experiments with diffusion at home (after all, you've freed up a lot of time now that you're no longer checking for news updates or watching video clips on Instagram). For example, diffusion can be used to make wilted or soft carrots crispy again.

HERE'S HOW IT WORKS:

- Fill a tall glass with water and place wilted carrots in it.

- Place the glass in the fridge and wait for a day (or a maximum of two days).

- The carrots will once again be crisp.

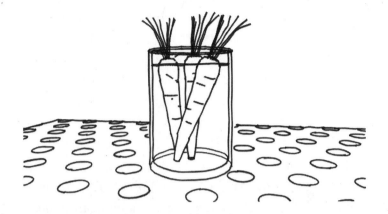

The process that plays out here is the same one that affected Wigald Boning's cell phone in the helium—only in this case it has been done on purpose. The carrots had already dried out and wilted. In the glass they were surrounded by a lot of water. Some of it "diffused" into the carrots and turned them crisp. By the way, when the carrots absorb water they also expand a little bit, so make sure to use a large enough glass; otherwise, you might not be able to pull the carrots out after they've been refreshed.

Cooking with osmosis

Unfortunately, we're waiting for two things now: the cell phone and the carrots. The wheels of physics turn slowly. Restless, we wander through the house and whenever we're bored, our thoughts turn to food. What could we cook quickly? There's a jar of preserved sausages in our cellar. We place the sausages in a pot, fill it with water, and turn on the stove. Big mistake! The sausages burst and no longer look appetizing. Thanks, physics!

What we failed to remember is that diffusion also plays a role in cooking—more specifically, an extension of diffusion: osmosis. In this process, water permeates a partially permeable layer, which allows some materials to enter and prevents others from doing so (scientists speak of a semi-permeable membrane). Imagine a sieve that children use in a sandbox to separate small pebbles from the sand!

It turns out that the skin of our sausages is semipermeable. It lets water in but keeps salt out. The sausages contain a lot of salt, but the water surrounding them does not. The sausages can't expel any salt because it can't pass through the skin. Consequently, there's only one option to balance out the salt–water concentration differential: The sausages must absorb more water. And that's exactly what happens; water permeates the sausages until the skin bursts. If only we had heated the sausages in salt water or in the liquid in the jar. If there's an equal amount of salt in the water and in the sausage filling, then there is no osmosis.

You may not taste the difference between a perfectly heated and a burst sausage, so let's look at another, somewhat fancier, example: boiled beef! If you want to prepare a particularly tasty piece of beef, the amount of salt in the boiling liquid should be as close as possible to the amount of salt in the meat to prevent any loss of flavor. It's the opposite when you cook a broth from scratch. In this case, you want to ensure that the flavor (and the marrow) of the bones dissipates into the water. And when you cook pasta, it's different again: It's best to add salt to the water so that the little salt contained in the noodles doesn't dissipate.

Dressing should only be added to lettuce immediately before serving as it wilts if covered in dressing for too long (the lettuce will happily diffuse the water it contains into the salty dressing).

Wrinkled hands

As we're washing the pot, we look at our hands in the sink. Our fingertips are all shriveled and wrinkly.

You'll often read that this is due to osmosis. The theory is that the body contains more soluble salt than the water. This is why water permeates our skin cells, causing them to swell, especially in areas where there are calluses. But this explanation isn't logical; in that case, our whole body would turn wrinkly, not just our fingers and toes. And to be precise, our fingertips don't really look swollen; on the contrary, they seem to be shriveled!

Scientists uncovered another explanation when they noticed that people with nerve damage can bathe for as long as they want without their fingers becoming wrinkly. Therefore, it must have something to do with nerves! The current state of knowledge is as follows: When we are in contact with water for long periods of time, the sympathetic nervous system causes the blood vessels in our extremities—toes and fingertips—to contract. In other words, the skin is retracted.[36]

---

36 To learn more about wrinkly fingers and possible reasons why they happen, see Richard Gray, "The Surprising Benefits of Fingers That Wrinkle in Water," BBC, June 20, 2022, https://www.bbc.com/future/article/20220620-why-humans-evolved-to-have-fingers-that-wrinkle-in-the-bath.

Boat sunk by osmosis

Sailors are likely the people who hate osmosis most passionately. Osmosis can sink boats, and it's hardly surprising that sailors speak of it as though it were a grave illness. A boat "has osmosis" or "is affected by osmosis." This specifically affects the hull, which lies beneath the water line. Older boats are often made of FRP (fiber-reinforced plastic) containing resin that isn't water-resistant over the long term. This means water can diffuse into the hull, where it collects in small voids that are always present in the laminate. The resin breaks down and forms an acid, which—in its effort to dilute—draws more moisture into the void. The liquid presses the boat's gelcoat outward, visible as blisters. When the blisters burst, the laminate is exposed to seawater without protection. This results in more and more damage occurring. If no one notices, the boat will eventually sink.

Damage of this kind has indeed caused loss of life—usually in the case of boats that have been immersed in water for years. Sailors who only go out on the water in spring and summer and pull their boats out onto a dry dock in winter have fewer problems (at least they have a chance to spot any osmosis damage when the hull is out of the water). Anyone planning a major sailing trip should buy a more modern boat, where different resins will have been used that are more resistant to water and to osmosis. A detail that's good to know!

In the meantime, Wigald Boning has forgiven us for the mishap with his iPhone. He didn't wait until the helium dissipated from his cell phone; he went to an Apple store with his broken cell phone and asked them to check it. The diagnosis: water damage. They gave him a new one.

IMPACT SCALE

Annoying

Lifehack

Catastrophic

What a Beautiful Glow!

We are constantly exposed to radioactive radiation and even eat food laden with radioactivity. Is this dangerous?

We don't really like to tell the story that follows, but if it must be told somewhere, then it should be in a chapter on radioactivity. After all, this is a story about a nuclear accident, more precisely about a self-inflicted accident.

It happened on a laboratory day at the end of our final semester. Lectures, assignments, and internships had all been completed; all that remained was the final dissertation. Marcus's topic was medical physics and whether radioactive implants could treat ocular tumors. The principal components of this experiment were gamma seeds—small radioactive capsules roughly the size of a grain of rice. They are composed of a titanium shell into which radioactive iodine-125 beads are embedded. As the iodine-125 decays, gamma radiation is generated; this can do severe damage to the human body.

In order to measure the radiation, these radioactive grains of rice have to be glued onto a synthetic block, naturally in accordance with stringent safety precautions that transform the whole operation into a game of dexterity. Marcus and the work surface were separated by a wall of lead bricks around forty centimeters (sixteen inches) high surmounted by a lead glass pane as a slanted roof. Marcus had to reach around this pane—not with his bare hands, of course, but wearing thick lead-protection gloves in which he held two long tweezers. Imagine the popular childhood game in which you don thick gloves, a scarf, and a beanie and try to unwrap and eat a bar of chocolate with a knife and fork. A race to eat chocolate with radioactivity thrown in.

And just like in the game, at some point you lose—in this case it was a radioactive bead. It simply slipped out of the tweezers' grasp and oops, it was gone. Without exaggeration, this was the worst moment of Marcus's entire time studying physics.

How do you find a tiny piece, millimeters in size and inconspicuously silver, when you have to wear protective goggles and bend over lead walls? Just searching seemed hopeless. Luckily logic kicked in after the initial shock: What do radioactive materials do? They radiate! Marcus went to get a Geiger counter and moved it slowly across the work surface. In one corner, the Geiger counter clearly showed more radioactivity than in the others. And there it was, at the very edge of the work area, the lost titanium capsule!

If you're thinking "Thank goodness I never studied physics, so I never had to work with radiation," we must disappoint you. Radioactive radiation is part of our environment, and we are all exposed to it—when we walk on cobblestones, fly on vacation, or eat bananas. That's right, bananas. They contain not only radioactive potassium but also alcohol, which makes us wonder why bananas are the world's most popular fruit (or maybe that's why they're so popular?).

Why does a banana contain radioactive material while an apple does not? This is linked to the atomic nuclei contained in both types of fruit (atomic cores, not apple cores!). Most atoms in our environment have stable nuclei. They remain stable regardless of whether they are moving through the air, encased within a cell phone battery, or hidden inside an apple. These stable atomic nuclei don't care which chemical compound the entire atom forms.

However, certain types of atomic nuclei spontaneously decay and are then called radionuclides (radio "cores"). The decay generates high-energy rays or particles, which spread at high velocity. This *ionizing radiation* is capable of ripping electrons from other atoms or molecules and chemically altering them. There are several escalating steps, depending on how much power the radiation has and how much damage it can inflict.

1. *Alpha radiation,* for example, occurs when the gas radon (which occurs in air) decays. This process creates helium nuclei that have a double positive charge, although they don't have a large reach in air. A sheet of paper suffices for protection. For this reason, alpha

radiation is relatively benign for humans unless the radioactive substance is inhaled or somehow absorbed directly into the body by another means.

2. *Beta radiation*: When naturally present potassium-40 decays to calcium-40, an electron escapes from the atomic nucleus. It has a considerably larger range than the helium nuclei—first because it is much smaller, and second because the electron initially moves nearly at the speed of light. To protect against beta radiation, more solid defenses are needed than paper. This calls for capsules made of metal; lead is best.

3. *Gamma radiation* is electromagnetic radiation. You could describe it as the big sister of uv light, although the wavelength is much shorter, and the radiation has much more energy. Gamma radiation is the ionizing radiation with the greatest reach. It is created during radioactive decay when a stimulated atomic nucleus changes state and releases energy. This is the radiation we were working with in the laboratory when the radioactive grain of rice was lost.

Gamma radiation penetrates most materials with ease, among them the human body. This is what makes it possible for us to be X-rayed when we break a leg, for example (although it should be noted that in an X-ray machine the gamma radiation is produced with the aid of high voltage, which is why no radioactive material is needed inside the machine).

Today we know that X-rays should be used sparingly. When we were children, attitudes were still a little different.

For example, we always had our feet X-rayed when we went shoe shopping to check whether the shoes were the right fit. Does that seem unbelievable? Well, it really did happen: In the 1970s, we would stand in front of a wooden box-like structure and place our feet inside the opening at the front of the box.

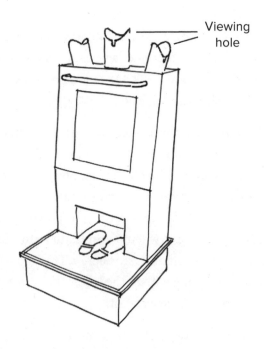

Looking through the viewing hole at the top, you would see a miracle of physics: an X-ray view of our feet. *Live.* X-ray light—that is, gamma radiation—was radiated through our feet and projected onto a kind of "TV screen" directly above our feet. This glass plate, which shone in a greenish light, showed how well our feet fit inside the new shoes we were trying on and whether we could still wiggle our toes. Of course, we children always tried on as many shoes as we could. Our feet were exposed to a fair amount of radiation—and today we shudder at the thought of the sales personnel, who stood next to the apparatus all day long without any protective clothing as they served customers. At the time, however, this was all the rage!

Even without X-raying our feet, we are regularly exposed to ionizing radiation. It's in the air, in stones and rocks, in our food; it even comes from the sky. Although the sources are so varied, physics allows us to draw comparisons between them all. And so the scale of radiation exposure is measured in units of sievert. This unit also indicates an extremely important value that we are going to reveal to you now. We can use this value to express the degree of danger posed by different sources of radiation. It is the *average natural radiation exposure* caused by ionizing radiation. This is 2.1 msv per year (msv = millisieverts) in Germany and 3.1 msv per year in the United States.[37] In this context, "natural" means that there hasn't been a Chernobyl reactor disaster, or you haven't just had an X-ray. That kind of exposure would be added on.

Where does natural ionizing radiation come from? Roughly half of it comes from the air, where the radioactive gas radon is located. The rest is found in terrestrial radiation through various minerals, radioactive elements in our food, and cosmic radiation.

---

37 "Radiation Sources and Doses," United States Environmental Protection Agency, last updated February 22, 2024, https://www.epa.gov/radiation/radiation-sources-and-doses.

Radiation exposure of 2.1 msv per year is of course a mean value. There will be significant variations depending on how and where you live. For example, southern Germany has a much higher radon concentration than the far north of the country; this is because of different types of rock from which radon emanates to the surface. If you beautified your flooring entirely with granite slabs, you would slightly increase the radiation exposure in your home because granite contains several radionuclides, including a tiny amount of uranium. If you love Brazil nuts, you will ingest the important trace element selenium, as well as some radionuclides. If you live high up in the mountains, you are exposed to more cosmic radiation than if you were standing on a dike by the North Sea.

We have summarized several radiation sources in the following tables, sort of like calorie tables for radioactivity:[38]

Radiation source	Annual exposure in msv
Radon	1.1
Terrestrial radiation	0.4
Food	0.3
Cosmic radiation	0.3
Sum of natural radiation	**2.1**

---

38 German Federal Ministry for the Environment, Nature Conservation, Nuclear Safety and Consumer Protection (BMUV), annual report on environmental radioactivity and radiation exposure (2018).

Event	Exposure in msv per event
Eating a banana	0.0001
Walking on cobblestones for one hour	0.0002
Eating a Brazil nut	0.0004
X-ray of an arm	0.005
X-ray of a lung	0.02
CT scan of the torso	8
X-ray during procedure to open coronary arteries	15–20
Return flight Frankfurt–New York	0.1
Half a year on the ISS	120
Threshold for safe exposure for medical personnel per year	20

As you can see, it's possible to measure the ionizing radiation of bananas. One banana equals 0.0001 millisieverts. This is one twenty-one-thousandth of the average annual exposure. Apparently, Geiger counters have been triggered in US ports when shipments of bananas have arrived.

If you're asking yourself how many bananas you can safely eat without getting cancer, we have good news for you: You can eat as many as you like! The radioactive potassium is excreted, so it doesn't stay in your body. Scientists have nevertheless (and perhaps for fun) coined the term "banana equivalent dose." We can compare other everyday radioactive sources using the banana equivalent (= 0.0001 millisieverts). One hour of walking on cobblestones equals two bananas; a round-trip flight between the US and Germany equals up to 1,000 bananas.

Thank you, cosmic radiation

Some 3,000 bananas emanate down from space each year in the form of cosmic radiation. While this may sound like a superhero scenario, it's actually a stream of protons and helium nuclei, the alpha particles. When these particles come into contact with the Earth's atmosphere, they collide with air molecules, accelerating them. This, in turn, creates entire showers of additional particles. However, only a few of these reach the Earth's surface. The remainder continue to swirl around space. Cosmic radiation is therefore 800 times stronger on board the ISS than on the ground.

Although cosmic radiation only plays a small role in everyday life, we should be thankful for it. Scientists surmise that it may have contributed to the formation of life on Earth. For complex chemical compounds to form, there must be energy and a certain chaos, to which cosmic radiation may well have contributed.

Even in lower layers of the Earth's atmosphere, cosmic radiation also ensures the presence of electrically charged particles, which come into play during thunderstorms. The high voltages in storm clouds can only discharge when there are sufficient moving charges: Voilà! Since lightning is also said to have played a role in the evolution of life on Earth, we can see once again that cosmic radiation is pretty cool.

Dangerous? Yes—but!

Is natural radioactive radiation dangerous? In brief, "yes" followed by an emphatic "BUT"! Let's get to the "yes" first: The danger of ionizing radiation is that it changes molecules. When the radiation is directed at a cell, it's possible that this cell will die, that it will no longer be able to propagate, or that its genetic material will be altered. The last may lead to cancer; theoretically, this could happen if just a single particle impacts us in an adverse way.

Let's quickly get to the "BUT." The probability of a dangerous alteration in genetic material is very, very small. Excess mortality resulting from natural ionizing radiation is difficult to calculate because it's impossible to determine after the fact whether cancer was caused by chemical exposure, viruses, radiation, or other external factors. Moreover, age, gender, and the affected organ also play a large role.

As obvious as it may be that frequent flying may increase the risk of cancer, it's equally difficult to prove it. Although a study has documented a higher risk for airline personnel,[39] it may also be that their irregular sleep schedules and chemicals in the cabin played a role in the statistical

---

39 Eileen McNeely, Irina Mordukhovich, Steven Staffa, Samuel Tideman, Sara Gale, and Brent Coull, "Cancer Prevalence Among Flight Attendants Compared to the General Population," *Environmental Health* 17 (2018): article 49, https://doi.org/10.1186/s12940-018-0396-8.

findings. Statistics are always a problem in the context of ionizing radiation. A flight to Mars would translate into a significant amount of radiation exposure for an astronaut. The concomitant cancer risk is nevertheless lower than the risk of a chain smoker dying because of their addiction.

Given these challenges, the consequences of low ionizing radiation over a long period of time can only be expressed as estimates.[40] Most of the data originate from survivors of the atomic bombs at Nagasaki and Hiroshima, observations during radiology tests and radiation treatments, and people whose work exposes them to radiation. The approximate results of these data are that natural ionizing radiation accounts for roughly 3–4 percent of the 230,000 people who die from cancer each year in Germany.

One more time: We are naturally exposed to this risk, whether we want to be or not. And it is very low. But the level of natural radiation exposure is a useful value upon which to decide how much additional radiation you want to be exposed to.

The discovery of radioactivity

Aside from horrific nuclear accidents such as in Fukushima, people have suffered from severe radiation damage when exposed to increased radioactive radiation. X-ray radiation and early radioactive elements were only discovered in the late nineteenth century by Antoine Henri Becquerel and

---

40 ICRP, "1990 Recommendations of the International Commission on Radiological Protection," ICRP Publication 60, Ann. ICRP 21 (1–3).

by Marie and Pierre Curie, but radiation had been causing injuries for a long time already. Back in the sixteenth century, the famous physician Paracelsus described the Schneeberg lung disease, which predominantly afflicted miners. They were mining iron ores, and also uranium, which had not yet been identified.

What does radiation do to us?

Does ionizing radiation kill cells? Most of the time it doesn't kill them directly. Instead, it lives up to its name by ionizing water or other components inside the cells. Radicals—that is, charged molecular particles—form and can cause a wide range of changes in the cell. In the worst (albeit unlikely) scenario, both strings of DNA are severed and thereby destroyed. When fewer radicals form in a cell, the body nearly always manages to repair the damage itself. If this fails, the cell may no longer be able to divide and will move toward cell death, or else it will mutate. The latter may then lead to the development of a tumor. Wherever cells reproduce at the greatest speed—for example, in the lining of the stomach—the risk of cancer resulting from radiation is at its highest.

Fortunately, many cancer cells struggle to recover from radiation exposure on their own. This is what doctors use when they treat tumors with radiation. The art is to expose the affected tissue to a high, yet calibrated, dosage so that the surrounding healthy cells can tolerate the treatment while the tumor cells cannot. This makes it possible to kill the tumor. Although the healthy tissue also suffers some damage, that damage will only surface later, and in the time that this gains, the patient can continue to enjoy a satisfactory quality of life.

Radiation therapy is mainly performed with special X-ray machines, although strongly accelerated electrons, protons, and other particles are also used. In the past, radioactive materials were also used; the radiation emanating from these materials was directed toward the treatment area. Of course, it was important to handle these radiation sources with great care, as they continued to be very dangerous long after use.

In 1987 (when the X-ray boxes in shoe stores had surely long been dismantled), two scrap collectors in Goiânia, Brazil, heard rumors that valuable equipment had been left behind on the site of a closed-down clinic. They searched the site and found a discarded X-ray machine, which they took apart with simple tools. They pulled out a valuable-looking metal cylinder, transported it home in a wheelbarrow, and placed it under a mango tree in the garden.

Within a day or two they both felt unwell. They were nauseous and listless. Doctors diagnosed an allergic reaction to spoiled food. The scrap collectors sold the metal cylinder to a friend who dealt in scrap metal.

The scrap dealer noticed a "pretty blue light" emanating from the cylinder. He broke up the cylinder with a hammer and crowbar to reveal a glowing stone, which he took into his house. He sold the metal to a farmer and a printer, among others. No one realized that what the scrap collectors had found was high-grade radioactive cesium-137 (beta radiation), which had been used to treat tumors.

For days, the cesium was cluelessly handed around, examined, and touched; a little girl rubbed the blue powder all over her arms. All fell ill or died. Nevertheless, it took two weeks before the wife of one of the victims suspected that the metal cylinder might be at fault. With the help of an acquaintance, she took it to a physician (on a public bus, in a shoulder bag). There it lay for some time, until the physician decided to consult a colleague who had knowledge of radioactive material. Together they alerted the authorities—and managed to return to the physician's home just in time to stop the fire service from disposing of the radioactive cylinder by throwing it into a river.

The accident had horrific consequences for the people of Goiânia.[41] Hundreds suffered from radiation exposure, several died, and increased radiation is measurable in the area to this day.

That's the problem with radioactive radiation: Nothing can be allowed to go wrong. If it does, then things really go wrong. Cesium in particular doesn't allow for even the slightest error in handling because it is highly chemically reactive and readily forms compounds with all kinds of materials. Even the relatively small amount of radioactive cesium that was released in Goiânia contaminated eighty-five homes. Several had to be demolished. In total, the incident resulted in 3,500 cubic meters (123,600 cubic feet) of contaminated radioactive waste, roughly equivalent to the load of 1,000 trucks. All this waste has to be

---

41 "Radiation Sources: Lessons From Goiânia," *IAEA Bulletin* 4/1988, https://www.iaea.org/sites/default/files/publications/magazines/bulletin/bull30-4/30402781017.pdf.

188 • THE UPS AND DOWNS OF PHYSICS

stored in a safe place for 180 years—the time it will take for the radiation that emanates from the waste to be low enough to be deemed harmless.

In principle, radioactive materials never stop radiating. Let's look at a single cesium-137 atom. It isn't stable, which means it will decay at some point. The probability of this happening can be determined by its half-life. For cesium-137, this is 30.17 years. This means there is a 50 percent chance that it will decay into stable and harmless barium-137 within 30.17 years, expelling an electron into the environment. But it might not change to a harmless state. And then it all starts up again. If you've ever played Ludo, you'll know that sometimes you have to roll the dice many times before you can move your piece to home base. In this case, it is the gods that roll the dice, if you will. For our cylinder filled with cesium-137, this means that half of the cesium will have decayed after 30.17 years, while the other half remains. After another 30.17 years, a quarter remains, and so on. In other words, the cesium-137 content halves every 30.17 years. The material will only become truly harmless when the radiation has become so weak that it falls within the range of natural radioactivity.

For a radioactive material, 30.17 years isn't all that long. Iodine-129 has a half-life of 17 million years before it decays. This is what makes it so difficult to find a truly safe permanent disposal site for nuclear waste. The quest is to find a location where nuclear waste can be stored for a million years. As humans, we can barely conceive of this timeframe.

Despite these dangers, radioactivity is an integral part of some fields, including space exploration. The

Perseverance Mars rover is equipped with a multi-mission radioisotope thermoelectric generator (MMRTG) that uses plutonium as its power source. The plutonium decays inside this device. This process generates a lot of heat, which is transformed into power through a thermal element and charges rechargeable lithium-ion batteries. There are very few other ways of ensuring an energy supply over a long period.

And then there are the people who voluntarily expose themselves to a bit of ionizing radiation—for example, in the form of a watch. Some watches contain a gradually decaying radioactive material. This creates alpha or beta radiation, which stimulates another fluorescent material so that the watch is always glowing—without batteries. This isn't a problem if the radiation remains inside the watch. And most of the time it does. But a problem arises with so-called *bremsstrahlung*. When fast electrons fly past atomic nuclei, they are strongly diverted. This creates radiation, which can escape and hit whoever is wearing the watch. However, the bremsstrahlung in a watch is extremely low: 0.02 msv per year, one hundredth of the naturally occurring exposure. Or just 200 bananas.

IMPACT SCALE

Annoying	🌧 🌧 🌧 🌧 🌧
Lifehack	💡 💡 💡 💡 💡
Catastrophic	💣 💣 💣 💣 💣

· FOR SMARTY PANTS ·

Do Radioactive Materials Really Glow?

If you're familiar with Homer Simpson, the cartoon character who works at the nuclear plant in Springfield, you'll know that radioactive materials glow green. Unfortunately, Homer isn't very smart. Otherwise, he would know that the green glow nearly always emanates from a radioactive luminous paint. This type of paint is used on the faces of some watches (see above)—but not in nuclear reactors.

Nuclear reactors glow blue

The Simpsons ought to show the ghostly blue light of fuel rods in a water-filled reactor pool. This light does exist. It's created by the *Cherenkov effect*, which occurs when a charged particle moves faster in a non-conductive medium than light does in the same medium.

In water, for example, the light is repeatedly scattered by particles in the water, and as a result it spreads 25 percent slower than in a vacuum. That is still a great speed. But the electrons that escape from the nuclear reactor are even faster. Indeed, they are nearly as fast as light would be in a vacuum: almost 300,000 km/s (186,411 mi/s).

When the electrons move through the water at this breakneck speed, they temporarily displace the electric charges in the water. This charge displacement generates weak electromagnetic radiation, which spreads in all directions. These electromagnetic waves overlap to form a cone-shaped wave front that drags behind the electron. This is Cherenkov radiation.

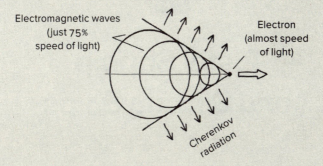

The wave front is reminiscent of the V-shaped wake of a swimming duck or a ship. There are more shorter wavelengths than longer wavelengths in Cherenkov radiation, hence the bluish glow.

In the nuclear accident in Goiânia, witnesses described the radioactive material as having a bluish glow. Presumably, the Cherenkov effect was at play here as well, because when cesium-137 decays, fast electrons are created. The material itself is translucent and the light moves within it at a speed that is two fifths slower than in a vacuum.

Because Cherenkov light spreads widely, it is particularly useful for observing rare, very-high-energy elementary particles. In the "IceCube" experiment, 5,160 light sensors were inserted deep into the Antarctic ice. This is a method of proving the presence of cosmic neutrinos, which create flashes of light. And those are blue as well. Not green.

Pistol Shrimp and Sinking Ships

Why cavitation is disruptive at sea and helpful in the kitchen

HMS *Daring* was to be the "Fastest Boat Ever." The British Royal Navy had outfitted it with an extra-large propeller, two water-tube boilers, and two steam engines.[42] At the time—it was 1893—this design was high-tech. A large torpedo was installed at the bow to immediately destroy any enemies.

At least in theory. In practice, the extra-large propeller churned at high speed, but the ship traveled slowly. No matter how much they revved the engine, the super ship moved across the sea at a snail's pace. The engineers spent a long time searching for the cause, but to no avail—until they finally looked beneath the hull. It looked like a whirlpool

---

42 Wikipedia, "HMS *Daring* (1893)," last edited July 27, 2023, 00:20 (UTC), https://en.wikipedia.org/wiki/HMS_Daring_(1893).

down there. Vast amounts of bubbles whirled and swirled around the propeller. This seemed odd to the engineers: Where did these bubbles come from? After all, the propeller was below the waterline, separated from the air above.

The British shipbuilders had stumbled across a physics phenomenon that, while extremely annoying for them, is also very interesting: cavitation. Derived from Latin, the term means "hollow space." And it was such hollow spaces that prevented the "Fastest Boat Ever" from fulfilling its promise to *be* the fastest boat ever.

So, what does cavitation do? Whenever objects slide through a liquid at speed, negative pressure is created behind them. You may know the feeling of standing close to a busy road as a truck passes by at high speed. Apart from getting a terrible fright, you feel like you're being sucked into a maelstrom. This is because the air in the wake of the truck moves at extremely high speed. For a fraction of time, there is too little air behind the truck. And where there is too little air, there is little air pressure; the pressure at this point is lower, and to compensate, the air in the immediate surroundings quickly flows in to fill the gap.

The same thing happens when a ship's propeller cuts through water—the only difference is that water is flowing, not air. The propeller blades create a powerful wake. Very fast currents occur, and the pressure drops behind the propeller.

This led to the formation of strange bubbles within which the propeller of the HMS *Daring* churned with such great effort. This was water vapor. Boiling water bubbles! This seems improbable at first glance because seawater is cold. However, the pressure behind the propeller has fallen sharply—and it is the ambient pressure that determines the temperature at which water begins to boil.

The Earth's normal air pressure is 1,013 millibars (mbar). At this pressure, as we all know, water reaches the boiling point at 212°F (100°C). When we lower the pressure, the boiling point temperature also drops. On Mount Everest, air pressure is just 325 mbar and water reaches the boiling point at 158°F (70°C). Numerous physics publications have explored the question of why you can't boil eggs on Mount Everest. The answer is that, although the water bubbles and boils, the temperature at the boiling point is too low. Should you ever want to climb Mount Everest, you'd do well to bring your breakfast with you from base camp. But you won't have to lug your beverages up the mountain: The boiling temperature up there is still sufficient for a cup of green tea.

You can observe this effect yourself by getting a syringe from a pharmacy and performing the following experiment.

YOU WILL NEED:

- A disposable syringe, preferably somewhat thicker and with a stopper at the opening

- Lukewarm water

HERE'S HOW IT WORKS:

- Fill the syringe to roughly one third with lukewarm water.

- Close the opening with the stopper (if there's no stopper or plug, use your finger or some Plasticine).

- Now pull the plunger of the syringe firmly, as if you wanted to pull more liquid into the syringe. This creates negative pressure. The water begins to boil.

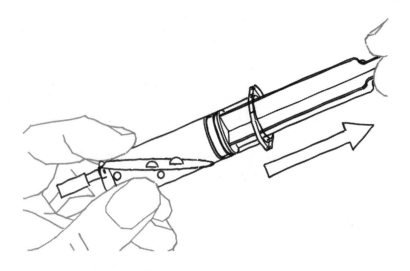

But this isn't about Mount Everest; this is about the waters in which the "Fastest Boat Ever" was floating in 1893. What about the pressure there? As all divers know, there is high pressure in the depths of the ocean. The more water there is above them, the higher the pressure. But the ship's propellers are just below the water's surface, where the pressure is not particularly great, and the boiling temperature of the water is correspondingly low. Because of cavitation, the pressure drops even more. And so the water behind the propeller does indeed come to a boil, resulting

196 • THE UPS AND DOWNS OF PHYSICS

in many small vapor bubbles. Physicists call them cavitation bubbles.

These bubbles don't last for very long because of the pressure the water exerts on them. And that's not all; the air is also exerting pressure on the surface of the sea. Within milliseconds, these bubbles implode—and then things get interesting! When such a small, round bubble implodes, all the forces of the current and the water surrounding it concentrate on a single, tiny point, the point that had been the center of the bubble. This is a highly unusual scenario, one that rarely happens in nature. And it releases enormous forces!

When the bubble bursts, small and incredibly fast currents form, known as microjets. Although they are tiny, they push forward with great force—imagine countless needle pricks. A hard, sharp needle can do a lot of damage precisely because all the force is concentrated on a single point. If you push long enough, a needle can prick and destroy a basketball, a wooden board, or even metal... The blades of large ships look quite beaten up when they've been exposed to the forces of cavitation over an extended period. The metal is peeling and dented, so it's easy to imagine how it might have been pummeled by needles.

The loudest
animal in the world

What may be an impediment to shipping is very useful in the animal kingdom. Some animals use cavitation to hunt prey or to defend against predators. The best example is the pistol shrimp, a five-centimeter-long (two-inch) crustacean. It is also rightly referred to as the "snapping shrimp" because it simply snaps at its foes. It uses its pistol (or snapping) claw to create a sound that is louder than the boom of a jet plane—200 decibels, which makes it the loudest animal in the world. Small animals faint, and larger predators quickly flee. Even submarine sonar can be disrupted by this pistol sound. And the shrimp achieves this with the help of cavitation. It snaps its pistol claw shut in an explosively fast movement, shooting a jet of water at its enemy. And behind this jet of water, a water-vapor-filled bubble forms (just as described above) then implodes, making the loud pistol sound.

What's amusing is that the pistol shrimp might not even hear this itself. Researchers have not been able to find any hearing organs in the shrimp, which may be a blessing considering the loud bang they make.

Moreover, the shrimp produces not only a thunderous sound with its claw, but also a flash of light: When the cavitation bubble implodes, so much energy is released that sonoluminescence is produced. This is a light effect created when liquids are subjected to strong pressure differentials. Unfortunately, as humans we cannot see the lightning with the naked eye. But when this super shrimp is filmed in exaggerated slow motion, the sonoluminescence becomes visible. It really is a sight to behold! The scientists who discovered this effect were so delighted that they even gave it the nickname "Shrimpoluminescence."[43]

---

43 Detlef Lohse, Barbara Schmitz, and Michel Versluis, "Snapping Shrimp Make Flashing Bubbles," *Nature* 413 (2001): 477–478, https://doi. org/10.1038/35097152.

The pistol shrimp is such a fascinating creature that it would be easy to write an entire book on the subject. In its private life, it is a very social creature, happily cohabiting with small fish or sea anemones. It frequently dwells in a cave together with a guardian goby, a small striped fish. The shrimp excavates the cave all day long and the little goby swims back and forth at the entrance keeping watch for enemies. Should an octopus come their way, for example, the little goby quickly retreats into the cave, trembling with fear. This is a signal to the shrimp, which rushes out of the cave and shoots at the aggressor with its pistol claw.

And should the shrimp be defeated and lose its pistol claw, it simply repairs itself; the normal claw on the opposite side is reconfigured into a pistol while a new pistol claw gradually regrows on the injured side.

Finally, a fast ship

Naturally, a ship like HMS *Daring* can't self-repair in such a spectacular way. In 1893, engineers needed to find a solution for the bothersome cavitation. And they succeeded! Instead of a single large propeller rotating at great speed, the ship was equipped with several smaller, slightly less powerful propellers. The immediate effect was that the water flow slowed a little and there was less cavitation. In the end, the ship reached a speed of thirty-two knots (59 km/h, or 37 mph), which was truly fast for the late nineteenth century. Newspapers finally wrote articles on the "Fastest Boat Ever" and the engineers were content.

Later on, they had to make a small adjustment. They had installed a torpedo tube at the bow to fire at enemy vessels.

This turned out to be impractical because HMS *Daring* was now so fast that it ran the risk of overtaking its own torpedo.

This all happened more than a century ago, and many sophisticated solutions have been developed in the meantime to protect ships against cavitation damage. Some ship propellers are constructed in such a way that air escapes along their edges. In this design, the many small air bubbles function like a damper. If the water flows too quickly behind the propeller and the pressure drops too much, they expand and thus prevent the formation of cavitation bubbles. This technique is also useful for warships because they render the ships quieter. As we learned from the pistol shrimp, imploding bubbles are quite noisy. And too much noise can cause a ship to be spotted by enemy sonar.

Cooking better with cavitation: Let physics work for you!

Even if you don't happen to be in the navy or own a large ship, you can still make good use of cavitation and even have a lot of fun with it. While cooking, for example! At some point in our lives, we've all stood in the kitchen, desperately trying to open a jar of pickles (or olives, or any other preserves). It doesn't matter whether we have sweaty hands or too little strength, the lid is tightly closed and cannot be unscrewed. Friends of ours have a marvelous tool in their kitchen for this problem, pliers that you can use to grip and unscrew the lid with the help of leverage.

Unfortunately, we don't have a tool like this, so we need to resort to the method used by Judith's Grandma Anni: Turn the jar of pickles upside down and enthusiastically tap the bottom of the jar with a flat hand. Grandma Anni and Grandpa Heinz had an allotment with a lot of fruit and vegetables. Every summer, the harvest would be preserved, huge jars full of cherries, red beets, or pumpkin. Whenever Grandma Anni was cooking, the characteristic slapping sound of her tapping on jars of preserves would be heard from the kitchen—followed shortly after by the "pop" when the lid finally yielded and could be unscrewed.

Many people know that this method works, but they don't know why. The general assumption is that by slapping the bottom of the jar we exert pressure on the glass, the glass presses on the pickling liquid and this in turn exerts pressure on the lid, which releases the vacuum. But the experts (female physicists and house husbands alike) aren't sure. One reader of *Die Zeit*, the prominent German weekly newspaper, once passed this question on to the editors. In a brilliant column entitled "Is this true?" she asked: "I think this is a rumor, because I don't believe that the vacuum is released. But I'm happy to admit my error if someone competent can give me a plausible explanation."[44]

The editors got to work. They researched with manufacturers of screw tops and preserve jars. The manufacturers came back with three explanations. First, that the lid was stuck and could only be dislodged through the tapping. Second, that the pickles were pressing against the lid and allowing air to flow into the jar from the outside. And third, that tapping (on the bottom of the jar) releases oxygen into the pickling liquid, which reduces the negative pressure.

44 Christoph Drösser, "Eine Frage, drei Antworten," *Zeit Online*, September 13, 2006, https://www.zeit.de/stimmts/1997/1997_47_stimmts.

While all these explanations may sound plausible, they are wrong. What helps to loosen the lid is cavitation. Imploding bubbles form within the pickling liquid, just as they form when a pistol shrimp attacks. Because this isn't visible to the naked eye, we ran a test and filmed it in slow motion.

When we tap on the bottom of the jar, the jar moves downward at the speed of the tap. In doing so, you might say, it leaves the pickles and the liquid behind in the air because they are sluggish: When we tap the bottom, we don't fully dislodge them, and they remain at their old position for a fraction of a moment. It's a bit like the trick when you pull out a tablecloth really quickly and all the plates stay on the table.

The jar is now at the bottom, the liquid is not. This creates negative pressure at the underside of the lid for a very brief moment. The liquid begins to boil, so that bubbles form and burst within a millisecond—releasing all the forces that are also capable of denting a ship's propeller. We hear the "pop" sound, and then the jar can be opened easily.

This works whenever the contents of the jar are relatively liquid. You can't use the same method to open a jar filled with jelly. Jelly is too congealed. It sticks to the bottom of the jar. If you want to create cavitation with apple jelly, you'll need to shake the jar repeatedly until the jelly has dissolved completely. While the physics effect will then work, the jelly won't be very appetizing.

We therefore recommend that you try a different, more spectacular experiment: popping out the bottom of a bottle with the help of cavitation.

Experiment:
Popping out a bottle bottom

YOU WILL NEED:

- An empty glass bottle (mineral water, lemonade, or beer—it doesn't matter); a bottle with a fastener or stopper is ideal

- Some water

- A rubber hammer or a piece of wood (e.g., a log)

- A bucket

- A protective glove

- A bit of courage

HERE'S HOW IT WORKS:

- Fill the bottle with water to just below the top.

- Hold the bottle above a bucket.

- Hold the neck of the bottle in one hand (please wear the protective glove on this hand), and the hammer in the other hand.

- Bring the hammer down firmly on the opening of the bottle.

- The bottom of the bottle will pop out and the water will pour into the bucket.

- If it fails, try, try again, as the saying goes. At some point, it will work!

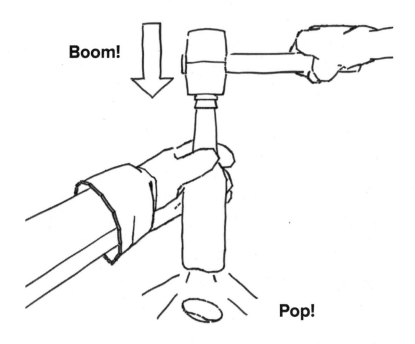

In this experiment, all you see at first is that the bottle is broken. But if you film it with a high-speed camera, you can see that several bubbles form along the bottom, and that these bubbles quickly implode and literally shoot out or pop out the glass bottom. The pistol shrimp sends its greetings!

There are videos online of strong men popping out the bottom of bottles by this means. The explanation is often that the pressure of their hammer blow is so strong that the air in the bottle presses down on the water, and that it is the water that then pops out the bottom. This isn't true—to achieve this, you would have to compress three liters (one hundred fluid ounces) of air into the bottle, which is very difficult to achieve. Besides, the trick also works when the bottle top is closed.

On the other hand, the same method can't be used to pop out the bottom of bottles filled with carbonated drinks. Just try it with a bottle filled with mineral water or lemonade: not a chance! The carbon dioxide in the beverage acts like a damper. It expands and ensures that the pressure doesn't fluctuate too much. And cavitation bubbles cannot form in the absence of negative pressure.

In principle, cavitation is possible wherever there's a lot of liquid. Even in the human body—after all, we consist mostly of water. And body cells don't tolerate little bubbles that suddenly implode with a lot of oomph. The diet industry tries to take advantage of this. Physicians and cosmeticians offer ultrasound treatments for all the cells their clients want to eliminate, like fat cells in the thighs, belly, and hips. The goal is to get these cells to pop in a controlled fashion. What remains is a fat-water mixture that is eliminated by the lymphatic system, then excreted or processed by the liver. We haven't researched how well this works. What is certain is that several sessions are required and that you need to budget over a thousand dollars for this treatment. Experts also caution that this treatment can only help in combination with changes in nutrition, massage, and exercise—all methods that tend to help with weight management without cavitation.

The effect of cavitation in medical therapy is more exciting. Cavitation can really help here—namely, in the treatment of cancer. Physicians target tumors with highly focused ultrasound waves. Cavitation bubbles form in the tissue, implode, and cut off the blood supply to the tumor.

This method is still relatively new, but research is ongoing, and it seems to be showing promising success already. And given this development, we can surely forgive cavitation the trouble it causes for ships.

IMPACT SCALE

Annoying

Lifehack

Catastrophic

Thank you!

WRITING A BOOK means being confronted with a fascinating aspect of physics: Time! And time, as we all know, is relative. At the beginning of a project, time seems abundant, an endless supply. Then it accelerates, passing faster and faster, until you have to speed things along, and there appears to be no time whatsoever for any other aspect of your life. For this reason, we extend our heartfelt thanks to Jannik, Swantje, Josephina, and Michel, who provided moral support, endless meals, and tolerance when conversation turned to physics, even at the dinner table.

Many thanks are also due to our agent Peter Molden and to Jessica Hein from Penguin Random House for the great care they provided and to Kanut Kirches and Stefan Heusler for their constructive and astute editing. On the scientific side, we received outstanding support from Svetlana Gutschank, Tobias Happe, Gerhard Heywang, Bernhard Niemann, and Thomas Seidensticker.

We are very grateful for the excellent and very patient service provided by Greystone Books, our publisher for the English-language edition.

It was pure luxury to write part of the book in very special conditions with a view of the Wadden Sea, with school lessons for our children projected right into our little island home during lockdown. Our profound thanks go to Margot and Joachim for making this possible. There can hardly be a more beautiful setting to deal with time pressure than a lonely beach overlooking the North Sea.

Index

aeolian harp (wind harp), 62

air: air pressure, 194; air quality, 156–57; flow resistance, 6–8, 9–11, 13, 75; impedance of, 142; ventilation, 155–56; weight of, 5–6, 160–61. *See also* humidity; wind

air conditioning, 153

alpha radiation, 176–77

alternating current (AC), 128–29

amplitude modulation, 49

Antarctic, 191

antistatic key chains, 123

Apple: iPhones and Watches, 163–65, 165–66, 167–68, 172–73

Aristotle, 68

atomic nuclei, 176

bananas, 176, 181–82

bath, hairdryers in, 127–28, 129–30

bathroom mirrors, 150–51, 159

Becquerel, Henri, 184

bees, 106–7

beta radiation, 177

bicycles. *See* cycling

blackboard, chalk on, 136

boats: HMS *Daring* and cavitation, 192–94, 195–96, 198–99; new designs, 199; osmosis and, 172

body. *See* human body

Boisjoly, Roger, 18–19, 20–21, 23–25, 27, 28–30, 31–33

Boning, Wigald, 162–63, 167, 172–73

boosters (solid fuel rockets), 19, 20–21, 23

bottles, popping out the bottom, 202–4

Brazil: Goiânia radiation accident, 186–88, 191

Brazil nuts, 180

bremsstrahlung, 189

Brewster, David, 116, 116n31

Brewster's angle, 116

bridges: Erasmus Bridge (Rotterdam, Netherlands), 63–64; Millennium Bridge (London), 55, 61, 64; in Ruhr region of Germany,

210 · THE UPS AND DOWNS OF PHYSICS

55–56; Tacoma Narrows
Bridge, 53–55
Brown, Robert, 167
Brownian motion, 167
burning glass effect: about,
102; concave facades on
high-rise buildings, 97–101,
105; concave mirrors, 99,
104–5; makeup mirrors,
101–2; reading glasses
experiment, 102–4; solar
power plants, 105
butterflies, 117

Cameroon, 77–78
cancer, 181–82, 183–84, 185–
86, 204–5
carbon dioxide, 77–78, 94, 156,
204
carpets, 119, 121, 123
carrots, 168–69
cars, 136–37, 138, 142–43
cavitation: boat design and,
199; cancer treatment and,
204–5; HMS *Daring* and,
192–94, 195–96, 198–99;
human body and, 204; loos-
ening jar lids and, 199–201;
pistol shrimp and, 196–97;
popping out a bottle bottom,
202–4
cellophane, 114–15, 116
cell phones: amplifying sounds
from, 139, 140, 141–43,
143–44, 145, 147; iPhones,
163–65, 165–66, 167–68,
172–73

cellular reception: capacity limits, 45; dead zones,
34–36, 46, 48; finding cell
towers near authors's home,
39–40, 40–41; 5G network,
41, 44, 45, 48–52; heat and,
46–48; physics of, 36–39,
41–42, 43–44; radiation and,
46; tower image problem, 46
center of the Earth, 78–79
centimeter waves, 44
cesium, 186–88
chalk, on a blackboard, 136
Challenger space shuttle disaster, 23–25, 27, 28–31, 31–32
Cherenkov effect, 190–91
China, 51–52
cleaning, 127
clothes, 27–28
clouds, 157
cocktail, gravity, 76–77
combustion point (focal point),
98–99, 104
concave facades, on high-rise
buildings, 97–101, 105
concave mirrors, 99, 101–2,
104–5
condensation, 152–53, 157–58,
158–60
conduction, 87–88
construction accident, 65–66
convection, 87–88
convex lenses, 102
cooking, 93, 153, 169–71,
199–201. *See also* cocktail,
gravity; food, in space
Copernicus, Nicolaus, 68

copper, 130
cosmic radiation, 182–83
crosswind, 11–12, 14, 15–17
Cuba, 3–5
Curie, Marie and Pierre, 185
cycling: aerodynamic bicycles
and positions, 12–14; crosswind and, 11–12, 14, 15–17;
in Cuba, 3–5; flow resistance
and, 6–8, 9–11, 13; head-
wind and, 4–5, 7, 9–10; to
North Sea, 8–10, 11; recumbent bicycles, 10–11; relative
wind and, 7–8, 10, 15–16;
slipstream and, 11; tailwind,
8–9; tips for, 14; Tour de
France, 10, 11, 12–13

dalton (Da, u), 161
dampers, 56–57, 64, 199, 204
Daring, HMS, 192–94, 195–96,
198–99
Darwin Awards, 72
dead zones. *See* cellular
reception
decimeter waves, 43–44
deserts, 152–53
diffusion: boats and, 172; carrots experiment, 168–69;
cooking and, 169–71; definition, 167; iPhones and
helium, 163–65, 165–66,
167–68, 172–73; wrinkled
hands and, 171
diopters, 103–4
dipole, Hertzian, 110–11, 112
direct current (DC), 128

Discovery space shuttle, 19,
20–21, 28
drag. *See* flow resistance

Earth: center of, 78–79;
Counter-Earth theory, 80;
greenhouse effect, 93–94
East Germany, 44–45
Edison, Thomas, 128–29
Einstein, Albert, 72–73, 167
elastomers, 21–23, 25
electricity: alternating current (AC), 128–29; hairdryer
in the bath and, 127–28,
129–30; lightning, 130–32,
132–33, 182–83. *See also*
electrostatic charges
electromagnetic radiation
(electromagnetism), 67n21,
72, 88–89, 107, 177, 190–91
electromagnetic waves (radio
waves): cellular reception
and, 36–39, 43–44, 45–46;
Cherenkov effect, 190–91;
heat and, 46–48; physical
barriers and, 41–42; types of,
42–43; "valley of the clueless"
and, 44–45
electrostatic charges: cleaning
and, 127; *Frag doch mal die
Maus* (Ask the mouse) quiz
show and, 124–26; friction
and, 121; Hindenburg disaster and, 133–34; human
body and, 119–20, 121–22;
laser printers and, 126–27;
static shocks, 119, 121–24

electrostatic induction, 133
Erasmus Bridge (Rotterdam, Netherlands), 63–64

falls, 67, 71–72
Feynman, Richard, 32
fingers, wrinkled, 171
fire, 1–2
5G cellular network, 41, 44, 45, 48–52
flooring, 109–11
flow resistance, 6–8, 9–11, 13, 75. *See also* wind
fluorine caoutchouc (FKM), 23
focal point (combustion point), 98–99, 104
foggy glasses, 149–50, 159–60
food, in space, 75–76. *See also* cocktail, gravity; cooking
Foster, Norman, 55
Frag doch mal die Maus (Ask the mouse; quiz show), 124–26
freezers, 26, 153–54
frequency, resonant, 61, 145–47
friction, 121
fridges, 153–54
fundamental forces, 67, 67n21

Galileo Galilei, 69
gamma radiation, 177
gamma seeds, 174–75
gases, greenhouse, 94
Germany: average natural radiation exposure in, 179; cellular reception and dead zones, 34–36, 40, 45, 48; "valley of the clueless," 44–45

glass. *See* greenhouse effect; windows
glasses: foggy, 149–50, 159–60; for reading, 102–4; sunglasses, 109
glass transition temperature (softening temperature), 25–26
Goiânia, Brazil, radiation accident, 186–88, 191
golf balls, 63
goosebumps, 135–36
granite, 180
gravitational force: introduction, 66–67; discovery of, 68, 69–70; equation (law of gravity), 70, 81–82; escaping gravity, 78–80; falls and mishaps, 65–66, 67, 71–72, 80–81; gravitational particles and waves, 72–73; gravity cocktail, 76–77; human bodies and, 70–71; Lake Nyos (Cameroon) and, 77–78; in space, 73–76
gravitons, 72
Greeks, ancient, 68
greenhouse effect: in children's rooms, 83–85, 87, 90–92; Earth and, 93–94; in greenhouses, 94–96; pasta sauce debacle and, 93; water vapor and, 94

hair, 119, 120
hairdryers, in bath, 127–28, 129–30
half-life, 188

INDEX · 213

hands, wrinkled, 171
headwind, 4–5, 7, 9–10
heartbeats, 129
heat: and cell phones and electromagnetic waves, 46–48; and conduction and convection, 87–88; heat radiation, 88–89; Planck's Law and, 89–90, 91. *See also* greenhouse effect; temperature
helium, 162–65, 166, 167–68, 172–73
Hertzian dipole, 110–11, 112
high-rise buildings, with concave facade, 97–101, 105
Hindenburg disaster, 133–34
Hooke, Robert, 69n23
human body: cavitation and, 204; electrostatic charges and, 119–20, 121–22; gravitational force and, 70–71; hairdryer in the bath and, 129–30; humidity created by, 151–52, 154–55; radiation effects, 185
humidity: bathroom mirrors and, 150–51, 159; benefits of, 157–58; comfortable levels, 151; foggy glasses and, 149–50, 159–60; freezers and, 153–54; from human bodies, 151–52, 154–55; insulation and, 157; static shocks and, 123; temperature and, 152–53; tips for preventing condensation, 158–60; ventilation and, 155; weight of dry vs. humid air, 160–61

IceCube experiment, 191
impedance, acoustic, 142, 143
induction, electrostatic, 133
infrared imaging cameras, 86–87
International Space Station (ISS), 73–76, 182
ionizing radiation, 176. *See also* radioactivity
iPhones, 163–65, 165–66, 167–68, 172–73
irregularly shaped bodies, 7

James Webb Space Telescope, 80
jars and jar lids, 19–20, 199–201
Journey to the Far Side of the Sun (movie, 1969), 80

Kármán vortex streets, 59, 60–61, 62, 63–64
Kepler, Johannes, 69n23
key chains, antistatic, 123
keys, 124

lactic acid, 116
Lagrange points, 79–80
Lake Nyos (Cameroon), 77–78
laser printers, 126–27
LeMond, Greg, 12–13
light, 85–87, 107–8, 108n30. *See also* burning glass effect; greenhouse effect; polarized light
lightning, 130–32, 132–33, 182–83
liquid crystal display (LCD) screens, 113–16

214 • THE UPS AND DOWNS OF PHYSICS

London: Millennium Bridge,
55, 61, 64; Walkie Talkie
(high-rise with concave
facade), 97–101, 105
long waves, 42
loudspeakers, 139–40, 142–43,
143–44, 147

makeup mirrors, 101–2
Mars, 82, 184
McAuliffe, Christa, 29
MEMS chips, 166
microjets, 196
microwaves, 41–42, 43, 46, 47
Millennium Bridge (London),
55, 61, 64
mirrors: bathroom, 150–51,
159; concave, 99, 101–2,
104–5; makeup, 101–2; parabolic, 104
Moisseiff, Leon Solomon, 54
mold, 154
Morton Thiokol, 19, 20, 23–24,
27, 28–29, 31, 32
Mount Everest, 194
mucins, 159–60

NASA: *Challenger* space shuttle
disaster, 23–25, 27, 28–31,
31–32; *Discovery* space
shuttle, 19, 20–21, 28
natural (resonant) frequency,
61, 145–47
Netherlands: Erasmus Bridge
(Rotterdam), 63–64
neutrinos, cosmic, 191
neutron stars, 82

Newton, Isaac, 68, 69–70
North Sea, cycling to, 8–10, 11

O-rings (sealing rings), 19–21,
23–25, 30–31, 32
oscillation, self-excited: dampers and, 56–57, 64; Erasmus
Bridge (Rotterdam), 63–64;
Kármán vortex streets
and, 59, 60–61, 62, 63–64;
Millennium Bridge (Lon-
don), 55, 61, 64; singing tea
strainers, 57–58, 60–62;
Tacoma Narrows Bridge,
53–55; washing machine
and, 56–57, 64; wind harp
(aeolian harp), 62
oscillators, 165–66, 167–68
osmosis, 170–71, 172. *See also*
diffusion

parabolic flight, 78
parabolic mirrors, 104
Paracelsus, 185
pasta sauce debacle, 93
Perseverance (Mars rover), 189
Pettenkofer, Max von, 156–57
phase modulation, 49
physics, 1–2, 33, 69–70. *See
also* air; burning glass effect;
cavitation; diffusion; electricity; electromagnetic
waves; electrostatic charges;
fundamental forces; gravitational force; greenhouse
effect; heat; humidity;
oscillation, self-excited;

polarized light; radioactivity; sound; temperature; wind
pistol shrimp, 196–98
Planck, Max (Planck's Law), 89–90, 91
plutonium, 189
podcasts, amplifying, 139, 142–43, 143–44, 147
polarized light: about, 107–8; bees and, 106–7; bluer skies and, 112–13; cellophane and, 116; flooring and, 109–11; home office experiments, 113–15; LCD screens and, 115–16; polarizing-filter film and, 108–9; Sara longwing and, 117
polyamide, 28
polyester, 27–28
polyethylene (PE), 26
polypropylene (PP), 26
power outlets, 127–28
pressure, 7, 137, 146–47, 193–96, 200–201
printers, laser, 126–27
Pythagorean theorem, 15

quantum physics, 90
quartz crystals, 166

radiation. *See* electromagnetic radiation; heat
radicals, 185
radioactivity: average natural radiation exposure, 179–80; bananas and, 176, 181–82;

cancer treatment and, 185–86; Cherenkov effect and bluish glow, 190–91; common sources, 180–81; cosmic radiation, 182–83; effects on human body, 185; Goiânia, Brazil, radiation accident, 186–88, 191; half-life, 188; increased exposure from human activities, 184–85; ionizing radiation, 176; missing gamma seed, 174–75; risks from natural exposure, 183–84; space exploration and, 188–89; types of radiation, 176–77; ubiquity of, 176, 179; in watches, 189; X-rays, 177–79
radionuclides, 176, 180
radio waves. *See* cellular reception; electromagnetic waves
radon, 176, 179, 180
rain radar, 44
recumbent bicycles, 10–11
refrigerators, 153–54
relative wind, 7–8, 10, 15–16
resistance, flow, 6–8, 9–11, 13, 75. *See also* wind
resonant (natural) frequency, 61, 145–47
rockets, solid fuel (boosters), 19, 20–21, 23
roofing accident, 65–66
Rotterdam (Netherlands): Erasmus Bridge, 63–64
rubber bands, 21–22

saccharic acid, 116

Sadler, Wendy, 101

salmon, on fire, 1–2

Sara longwing (*Heliconius sara*), 117

Sartre, Jean-Paul, 101

Schneeberg lung disease, 185

Scotch tape, 114–15, 116

screens, electronic, 113–16

sealing rings (O-rings), 19–21, 23–25, 30–31, 32

self-excited oscillation. *See* oscillation, self-excited

short waves, 43

shrimp, pistol, 196–98

sievert, 179

The Simpsons (TV show), 190

skin effect, 130–31

sky, blue, 112–13

Slinky, 141

slipstream, 11

smartphones. *See* cell phones

snakes, 86

softening temperature (glass transition temperature), 25–26

solar power plants, 105

solid fuel rockets (boosters), 19, 20–21, 23

sound: amplifying podcasts, 139, 142–43, 143–44, 147; chalk on a blackboard, 136; goosebump-inducing, 135–36; impedance and, 142, 143; loudspeakers, 139–40, 142–43; natural frequencies of mugs, 145–47; properties of, 141–42; thrumming

through open windows, 136–38; vases as loudspeakers, 143–44, 145, 147

space: *Challenger* space shuttle disaster, 23–25, 27, 28–31, 31–32; *Discovery* space shuttle, 19, 20–21, 28; food in, 75–76; International Space Station, 73–76, 182; Lagrange points, 79–80; *Perseverance* Mars rover, 189

spit, 159–60

static shocks, 119, 121–24. *See also* electrostatic charges

sticks, whipped through air, 58–59

stick-slip effect, 136

strong interaction (strong force), 67n21

Suhr, Wilfried, 61–62

sun, 79, 80

sunglasses, 109

sunlight, 85–87, 112–13. *See also* burning glass effect; greenhouse effect; polarized light

superposition, 39

synchronization, 55. *See also* oscillation, self-excited

synthetic materials, 27–28, 142–43. *See also* elastomers

Tacoma Narrows Bridge, 53–55

tailwind, 8–9

tea strainers, singing, 57–58, 60–62

temperature: boiling point, 194–95; elastomers and,

22–23, 25; glass transition temperature (softening temperature), 25–26; humidity and, 152–53. *See also* greenhouse effect; heat

Tesla, Nikola, 128

Thales of Miletus, 121

thermal imaging, 86–87

Timmer, Henning, 80–81

tires, 23

Tour de France, 10, 11, 12–13

transverse wave, 107–8, 108n30, 141n32

true wind, 7

ultrashort waves, 43

United States of America, 51–52, 67, 130, 179

uranium, 180, 185

Urmel fliegt ins All (Urmel flies to space; children's book), 80

vases, as loudspeakers, 143–44, 145, 147

ventilation, 155–56

Viñoly, Rafael, 99, 101

vortices: Kármán vortex streets, 59, 60–61, 62, 63–64

Walkie Talkie (high-rise with concave facade, London), 97–101, 105

washing machines, 56–57, 64

watches, 189. *See also* Apple

water: boiling point, 194–95; condensation, 152–53, 157–58, 158–60; greenhouse

effect and water vapor, 94; hairdryers in bath, 127–28, 129–30. *See also* cavitation; humidity

weak interaction, 67n21

Weber, Sebastian, 12

Westinghouse, George, 128

wind: calculating wind resistance, 16–17; crosswind, 11–12, 14, 15–17; flow resistance, 6–8, 9–11, 13, 75; headwind, 4–5, 7, 9–10; relative wind, 7–8, 10, 15–16; tailwind, 8–9; true wind, 7

wind harp (aeolian harp), 62

windows: greenhouse effect and, 84–85, 87, 90–93; thrumming through open windows, 136–38

wind turbines, 6

Wooldridge, Erik, 164

wrinkled hands, 171

X-rays, 177–79